Logarithmic Integral Equations in Electromagnetics

Logarithmic Integral Equations in Electromagnetics

Yu.V. Shestopalov, Yu.G. Smirnov
and E.V. Chernokozhin

///VSP///
UTRECHT • BOSTON • KÖLN • TOKYO
2000

VSP BV
P.O. Box 346
3700 AH Zeist
The Netherlands

Tel: +31 30 692 5790
Fax: +31 30 693 2081
vsppub@compuserve.com
www.vsppub.com

First published in 2000

ISBN 90-6764-322-X

Printed in The Netherlands by Ridderprint bv, Ridderkerk.

Contents

Introduction iii

1 Elements of the theory of integral operators 1
- 1.1 Integral operators with purely logarithmic kernel 1
- 1.2 Integral operators in Hölder spaces 5
- 1.3 Logarithmic integral operators and Chebyshev polynomials 14
- 1.4 Integral operators defined on a set of intervals 21
- 1.5 Integral operators with fixed logarithmic singularities 24
- 1.6 Elements of spectral theory 29
- 1.7 Abstract pole pencils 35
- 1.8 Logarithmic integral operators in Sobolev spaces 37
- 1.9 Integral operators with kernels represented by series 44
- 1.10 Methods of small parameter 46
  - 1.10.1 Approximate inversion 46
  - 1.10.2 Approximate semi-inversion 49

2 Generalized potentials with logarithmic kernels 55
- 2.1 Generalized potentials 55
- 2.2 Green's potentials 57
- 2.3 Examples for canonical domains 60
  - 2.3.1 Half-plane 60
  - 2.3.2 Rectangle 60
  - 2.3.3 Circle 61
  - 2.3.4 Exterior of a circle 62
  - 2.3.5 Ring 63

3 Summation operators 65
- 3.1 Matrix representation 65
- 3.2 Galerkin methods and basis of Chebyshev polynomials 67
- 3.3 Summation operators in the spaces of sequences 70
- 3.4 Matrix representation of logarithmic integral operators 73

4 Boundary value problems 79
- 4.1 Formulation of the problems 79
- 4.2 Uniqueness and existence theorems 82
- 4.3 Canonical problems: diffraction by strips and slots 87
  - 4.3.1 Diffraction by a slot 87
  - 4.3.2 Diffraction by a strip 89
  - 4.3.3 Diffraction by a screen with a rectangular slotted cavity 90

4.3.4 Scattering by a circular slotted cylinder 92
4.4 Eigenoscillations of open and closed slot resonators 96
4.4.1 Closed rectangular slot resonator 96
4.4.2 Open rectangular slot resonator 99
4.4.3 Slotted resonator with circular cross section 101
4.5 The integral and summation equations for the strip problems 104
4.6 Summation equations in the problem on eigenfrequencies 109

Bibliography **112**

Introduction

The purpose of this book is to give a review of basic results of the theory and applications of the integral operators and operator-valued functions with a logarithmic singularity of the kernel. Along with fundamental topics, we will present a self-review outlining some latest original results related to various applications.

Analysis of logarithmic integral operators is based on the theory of singular integral operators with the Cauchy kernel, which has been elaborated by Muskhelishvili, Vekua, and Gakhov in classical monographs [1]–[3]. A comprehensive study of singular integral operators is carried out in [4] from the viewpoint of the notion of a hypersingular operator. Monographs [3, 5, 6] address, among all, the questions of the index theory of the singular (Cauchy) and logarithmic operators. In [7], weakly-singular integral operators are considered in the L_p-spaces. A great variety of books, for example, [8]–[10], develop the solution technique for the integral equations with logarithmic kernels.

The trace theorems in the Sobolev space $H^1(\Omega)$ for a two-dimensional Lipschitz domain are known [11]–[14] to be well-established, as well as the potential theory in spaces $H^{\pm 1/2}(\partial\Omega)$ (see also [15]). The boundary and transmission problems for the Laplace, Helmholtz, and other elliptic differential operators in the two-dimensional Lipschitz domains, including their reduction to the boundary integral equations with logarithmic singularity of the kernel, have been studied, e.g., in [13, 16]. The latter paper focuses on the results concerning boundary integral equations with strongly elliptic Fredholm operators in the transmission boundary value problems for the Helmholtz equation in the Sobolev spaces; in addition, one can find a complete list of references to the works that study the properties of boundary integral operators with a logarithmic singularity of the kernel over a finite interval $(-a, a)$ in the Sobolev spaces.

Logarithmic integral operators usually arise in the so-called crack, slot, and strip problems. As far as the applications are concerned, such problems will occupy a major place in our studies. Note in this way works [17, 18], where the mixed boundary value (screen) problems for the Maxwell and Helmholtz equations were investigated. It is also worth indicating papers [19]–[21], where such problems are reduced to the summation equations with infinite-matrix operators.

In this book, we will study integral eqations with a logarithmic singularity of the kernel over a finite interval and over a set of such intervals; the slot and strip two-dimensional boundary value problems for the Helmholtz and Maxwell equations are reduced to such equations. A detailed consideration of the corresponding models connected with the electromagnetic and acoustic wave diffraction may be found in monograph [22]. Unlike the approaches developed, e.g., in [11, 13, 14], where the boundary integral operators are studied in the fraction-order Sobolev spaces, we will consider the integral operators with the logarithmic singularity of the kernel also in the weighted Hölder spaces that naturally appear due to the conditions imposed at irregular points of the boundary (edges). Such a choice of functional spaces enable one to precisely describe the properties of the logarithmic integral operators; namely, to establish their Fredholm property, and to separate the solution singularities close to edges.

In addition to the aforementioned papers, we indicate monograph [23], as well as [24] and some recent publications [25, 26]. In these works, the boundary logarithmic integral equations are reduced with the help of the semi-inversion based, in particular, on the method of the Riemann–Hilbert problem, to the coupled equations and to the equations with infinite-matrix (canonical Fredholm) operators with the trigonometric [27] and other types of infinite-matrix kernels acting in the wieghted l_2-spaces of the sequences of the Fourier-coefficients. The findings of Sologub [28] must be indicated: he has proved the unique solvability of the integral equation with the kernel in the form of the Hankel function, which corresponds to the classical problem of the plane electromagnetic wave diffraction by a flat strip. Note that a similar result was published almost simultaneously by Wolfe [20].

Among the books devoted to various applications of the boundary integral equations with the Cauchy and logarithmic kernels, we note monograph [9] that studies the elasticity problems, as well as [10], where the wave diffraction by unclosed cylindrical screens are considered.

Before to characterize the general contents of the book, we give some general remarks concerning important applications of logarithmic integral operators to solving boundary eigenvalue problems. We will study logarithmic integral operators with the kernels depending on one or several complex parameters. We develop specific methods for analyzing the unique solvability and problems on characteristic numbers (CNs) by considering these operators as operator-valued functions of appropriate complex (spectral) parameter(s). Such methods complement the known classical approaches. The corresponding operator equations are naturally connected with the two-dimensional boundary eigenvalue problems for the Helmholtz operator, where the spectral parameter enters the boundary conditions (and the conditions at infinity for unbounded domains) in a nonlinear way. In fact, any boundary eigenvalue (spectral) problem of mathematical physics, in particular those arising in electromagnetics, are reduced to operator eigenvalue problems nonlinear with respect to spectral parameter z,

$$K(z, \vec{a})\varphi = 0, \tag{0.1}$$

where $K(z, \vec{a}) : X \to Y$ is a multi-parameter operator-valued function acting on the pair of Banach spaces X, Y, and $\vec{a}$ is a vector of nonspectral parameters. Among them, there are parameters that describe the geometry of the domain, coefficients of the differential operator, etc. One may recall the simplest examples in the form of various transcendental equations, to which the Sturm–Liouville boundary eigenvalue problems on an interval are reduced. In the majority of applied problems, it is necessary to calculate families of vector-valued functions $z = z(\vec{a})$ determined implicitly by (0.1) rather than separate eigenvalues.

In electromagnetics and acoustics, spectral parameter z is usually related to the longitudinal wavenumber of a running wave or to the frequency of a resonant oscillation. (0.1) is often called the generalized dispersion equations, and the dependences of the CNs $z = z(\vec{a})$ on various nonspectral parameters are called generalized dispersion curves.

These problems were studied beginning from the early 1980s in [29, 30]; the results were summarized in monographs [24, 26, 31] (note also an important contribution made in [32, 33]) in connection with the modeling of the electromagnetic wave propagation and diffraction and analysis of resonance phenomena in open cylindrical waveguides and cavities.

This approach may seem to be impractical: indeed, initially linear eigenvalue problems are reduced to nonlinear functional-operator equations. However, its efficiency has been confirmed many times (see, for example, [26, 34]), especially when it is necessary to determined the dependence of eigenvalues on parameters; in perturbation theory; and, particularly, when K is an integral operator. In fact, a variety of special methods have been proposed that enable one to prove the existence and verify basic properties of the spectrum in terms of equation (0.1) and to efficiently calculate eigenvalues. Some of these methods will be considered below.

We note, in particular, that if K is a Fredholm operator, then its CNs form a countable set of (complex) points, and the spectral data $z(\vec{a}) = \{z_\nu(\vec{a})\}_{\nu=1,2,\dots}$ can be obtained in the form of numbered families. Often, it is convenient to vary one nonspectral parameter (a_0) which may be chosen simultaneously as a small parameter (for example, the diameter of an irregular perturbation of the boundary—a slot) and to obtain spectral data as asymptotic series with respect to a_0.

The typical example is K in the form of a scalar or matrix integral operator-valued function

$$K(z, a_0)\varphi \equiv \int_S K(t_0, t; z)\varphi(t)dt = 0, \quad t_0 \in S,$$

where there is only one nonspectral parameter, $a_0 = \operatorname{diam} S$. Another well-studied and typical situation is when K is defined by an infinite matrix $\|k_{mn}(z, \vec{a})\|_{m,n=1}^{\infty}$ and considered in the Hilbert space (l_2) of infinite complex sequences.

In many cases, K is a finite-meromorphic operator-valued function of a definite structure; namely,

$$K = K(z) = P_n(z) + U_n(z) = T(z) + \frac{A_n(z)}{z - \zeta_n} + U_n(z),$$

where $A_n(z)$ is a finite-dimensional operator; $T(z)$ is an invertible operator (which may involve an integral operator with a purely logarithmic kernel); and $U_n(z)$ is a completely continuous operator. In this case, CNs are located in the vicinities of poles ζ_n. A more detailed analysis of such operators is performed in Chapter 1.

The book consists of four chapters. Chapter 1 dwells on some basic methods of the theory of integral operators and operator-valued functions with a logarithmic singularity of the kernel. The approach is based on the generals methods of the spectral theory of operator-valued functions [35, 36] developed in [26, 31] for integral operator-valued functions. Note in this way that the methods [36] developed for three-dimensional problems, when the fundamental solution of the Helmholtz operator is analytic with respect to the frequency spectral parameter, cannot be directly applied in the two-dimensional case. The elaboration in [24]–[27] and [31] of the approach, which is different from the method [36] is caused, in particular, by the necessity of considering two-dimensional diffraction problems in unbounded domains. In fact, in order to determine complex eigenvalues, one cannot apply the standard Sommerfeld conditions at infinity, and it is necessary to construct an analytical continuation of the boundary operators and fundamental solution of the Helmholtz equation to an appropriate complex Riemannian manifold.

In Sections 1.1 and 1.2, we present the known formulas [1, 3, 4] for the inverse of the purely logarithmic and Cauchy operators and consider the unique solvability of the corresponding integral equations in the weighted Hölder and Sobolev spaces. In Section

1.2, we prove the unique solvability of the integral equation of the first kind applying, in particular, the original estimates [37]–[39] of the resolvent with respect to the norm in the weighted Hölder spaces. In Section 1.3, we apply the Chebyshev polynomials to the analysis of logarithmic and hypersingular integral equations. Section 1.4 extends the results of Sections 1.1 and 1.2 (using the technique proposed in [26, 37]) to the case when the set of integration consists of several intervals. In Section 1.5, we consider the properties of integral operators with fixed logarithmic singularities of the kernel. Section 1.6 plays an auxiliary role: it contains the references to the spectral theory of operators and operator-valued functions, including several important definitions and theorems. Integral operator-valued functions (with the kernels depending on a complex parameter) are considered separately; several examples are presented. Following [31, 39], we pay a special attention to the meromorphic dependence and the corresponding finite-meromorphic operator-valued functions. In Section 1.7, we consider a specific family of finite-meromorphic operator-valued functions called (abstract) pole pencils and derive explicit formulas for some inverse operators and CNs. In Section 1.8, we study logarithmic integral operators in the Sobolev spaces in terms of pseudodifferential operators (PDOs) using the relations between these operators and Chebyshev polynomials. In Section 1.9, we use the approach proposed in [39, 40] to prove the existence of CNs of the integral operator-valued functions with the kernels represented by series with meromorphic coefficients. Section 1.10 develops the methods of the perturbation theory as applied to approximate solution of logarithmic integral equations and determination of CNs. We perform a detailed analysis of the method of approximate inversion of integral operators with a logarithmic singularity of the kernel [42] and derive explicit formulas for inverse operators and CNs.

Chapter 2 is based on the results set forth in [37]–[39] and deals with definitions and essential properties of the generalized single- and double-layer potentials with logarithmic kernels and the Green's potentials. The potentials are considered in the two-dimensional domains with the boundaries containing a rectilinear interval. The corresponding theorems have been proved in [38]. Section 2.3 provides the reader with several explicit representations of the Green's functions of the Helmholtz operator for canonical domains, in which the logarithmic singularity is separated and the dependence on the parameter of the Helmholtz operator is analyzed.

In Chapter 3, following [25]–[27] and [31, 43, 44], we study fundamentals of the theory of infinite-matrix (or summation) operators and operator-valued functions in the Hilbert space with applications developed in [24, 26, 27, 44, 45] to the solution of logarithmic integral equations. Section 3.1 considers the matrix representation of an operator in the Hilbert space obtained by the general methods [43]. In Section 3.2, we describe the methods of approximate solution to linear operator equations and apply the general results concerning the substantiation and convergence of the Galerkin methods to solving logarithmic integral equations. In Section 3.3, we consider summation operators in the spaces of sequences. The definition of summation operators [44] are given and their specific family introduced in [44] (L-operators) are studied that are associated with logarithmic integral operators. In Section 3.4, following [44], we construct matrix representations of the integral operators and operator-valued functions with a logarithmic singularity of the kernel.

In Chapter 4, we consider some boundary value problems for the two-dimensional Helmholtz equation, including the cases of unbounded domains with noncompact boundaries (as applied to the wave diffraction, such problems were formulated as early as in

1940s, in particular, by Fel'd; the results are summarized in [46]). Section 4.1 presents the detailed formulations of the problems with the generalized condition at infinity introduced in [47] and applied in [31, 36, 48] to the determination of scattering frequencies, eigenoscillations, and eigenwaves. The analysis of solvability in terms of boundary integral equations with a logarithmic singularity of the kernel considered in the Sobolev spaces is performed in Section 4.2. Various canonical problems are considered in Sections 4.3 and 4.4. Integral equations with a logarithmic singularity of the kernel are applied, following [26, 39] and using the methods developed in [41, 42, 45], to the problems of the plane wave diffraction and determination of eigenfrequencies of slotted cavities and screens. The latter structures constitute an example of domains with irregularly perturbed nonperiodic infinite boundaries, where the standard methods cannot be directly applied. Note that such formulations have been first proposed in [49]. The Fredholm property and the uniqueness theorems are proved together with the discreteness of the spectra of eigenfrequencies (eiganvalues). In Section 4.3, we solve inhomogeneous boundary value problems formulated in Section 4.1 that simulate the wave diffraction by the considered structures. The main attention is paid to deriving explicit asymptotic representations of the solutions to canonical problems using the method of approximate inversion proposed in Section 1.9. In Section 4.4, we apply the methods of the spectral theory of integral operator-valued functions developed in Chapter 1 to the determination of eigenoscillations in two-dimensional electromagnetic resonators. We prove the existence of isolated complex eigenvalues with strictly negative imaginary part (scattering frequencies) for the problems in unbounded domains with noncompact nonperiodic boundary using the approach developed in [45, 49]. These problems are reduced to the determination of CNs of boundary Fredholm integral operator-valued functions with logarithmic kernels that are finite-meromorphic with respect to the frequency spectral parameter. The CNs of these operator-valued functions are determined by the perturbation technique justified in Section 1.9. The application of the method of summation equations produced by infinite matrices is considered in Sections 4.5 and 4.6. Section 4.5 occupies a special place in our study: it addresses the solution to the problem of diffraction by an arbitrary curved strip obtained in [25] using an alternative to the technique set forth in this book. This alternative method is based on the use of summation operators with trigonometric kernels. In Section 4.6, we again consider the canonical problem on eigenfrequencies of a rectangular slotted resonator using the method of summation equations constructed [25] in the basis of the Chebyshev polynomials. The problem is reduced to a canonical Fredholm infinite-matrix equation of the second kind.

Chapter 1

Elements of the theory of integral operators

In this Chapter, we present fundamentals of the methods applied to the study of the integral operators and operator-valued functions with a logarithmic singularity of the kernel.

We will employ the general definitions and notions of the spectral theory of operator-valued functions (the resolvent set, spectrum, CN, problem on CNs, finite-meromorphic and Fredholm operator-value functions, etc.). Section 1.6 contains all necessary references to the spectral theory of operators and operator-valued functions, including the definitions and theorems that we use in this Chapter. Note that a more detailed information can be found, for example, in the papers and books by Gohberg *et al.* [6] and [35], Sanchez-Palencia [36], [31], or in any textbook on the spectral theory.

1.1 Integral operator with the purely logarithmic kernel

We will begin the analysis of integral operators with a logarithmic singularity of the kernel (logarithmic integral operators) considering the simplest integral equation

$$L\varphi \equiv \frac{1}{\pi}\int_a^b \ln\frac{1}{|t_0 - t|}\varphi(t)dt = f(t_0), \quad t_0 \in [a,b], \tag{1.1}$$

with respect to unknown function φ for a given right-hand side f.

Assume that f is a differentiable function. Then, differentiating both sides of equation (1.1) with respect to t_0, we obtain the singular integral equation with the Cauchy kernel

$$S\varphi \equiv \frac{1}{\pi}\int_a^b \frac{\varphi(t)dt}{t - t_0} = f'(t_0), \tag{1.2}$$

where the integral on the left-hand side is understood in the sense of the principal value, i.e.,

$$\int_a^b \frac{\varphi(t)dt}{t - t_0} = \lim_{\varepsilon\to 0}\left(\int_a^{t_0-\varepsilon} + \int_{t_0+\varepsilon}^b\right)\frac{\varphi(t)dt}{t - t_0}.$$

It is natural to consider the integral equation with the Cauchy kernel in the class of functions satisfying the Hölder condition (see [1]).

Denote by $H_\mu(\Gamma)$ the set of complex functions φ defined on the segment $\Gamma = [a,b]$ and satisfying the Hölder condition with index μ $(0 < \mu \leq 1)$. This means that, for a given function φ, there exists a constant A such that, for any $t', t'' \in \Gamma$,

$$|\varphi(t') - \varphi(t'')| < A|t' - t''|^\mu.$$

The set $H_\mu(\Gamma)$ is a Banach space with the norm

$$\|\varphi\|_{H_\mu} = \|\varphi\|_C + h_\mu(\varphi),$$

where

$$\|\varphi\|_C = \sup_{t\in\Gamma} |\varphi(t)|,$$

and

$$h_\mu(\varphi) = \sup_{t',t''\in\Gamma} \frac{|\varphi(t') - \varphi(t'')|}{|t' - t''|^\mu}.$$

Denote by $\tilde{H}_\mu(\Gamma)$ the set of functions φ that can be represented in the form

$$\varphi(t) = \frac{\varphi^*(t)}{R(t)}, \tag{1.3}$$

where $\varphi^* \in H_\mu(\Gamma)$ and

$$R(t) = \sqrt{(b-t)(t-a)}. \tag{1.4}$$

The set $\tilde{H}_\mu(\Gamma)$ is a Banach space with the norm

$$\|\varphi\|_{\tilde{H}_\mu} = \|R\varphi\|_{H_\mu} = \|\varphi^*\|_{H_\mu}.$$

Denote by $\tilde{H}_{\mu,0}(\Gamma)$ the set of functions ϕ that can be represented in the form (1.3), where function φ^* equals zero at the ends of segment Γ. The set $\tilde{H}_{\mu,0}(\Gamma)$ is a closed subspace of $\tilde{H}_\mu(\Gamma)$.

Finally, denote by $\tilde{H}^1_{\mu,0}(\Gamma)$ the set of functions whose derivatives belong to $\tilde{H}_{\mu,0}(\Gamma)$. The norm in this Banach space is defined as

$$\|\varphi\|_{\tilde{H}^1_{\mu,0}} = \|\varphi\|_C + \|\varphi'\|_{\tilde{H}^1_\mu}.$$

Assume that function f' in equation (1.2) satisfies the Hölder condition. Then, according to [1], the general solution to (1.2) in the space $\tilde{H}_\mu(\Gamma)$ is given by the formula

$$\varphi(t) = L^{-1}f \equiv -\frac{1}{\pi}\int_a^b \frac{\sqrt{(b-t_0)(t_0-a)}f'(t_0)dt_0}{\sqrt{(b-t)(t-a)}(t_0-t)} + \frac{C}{\sqrt{(b-t)(t-a)}}, \tag{1.5}$$

where C is an arbitrary constant. This constant C must be chosen so that φ would satisfy equation (1.1). Substituting expression (1.5) into (1.1) and taking into account that

$$\frac{1}{\pi}\int_a^b \ln\frac{1}{|t_0-t|}\frac{dt}{\sqrt{(b-t)(t-a)}} = \ln\frac{4}{b-a}, \quad t_0 \in [a,b],$$

we obtain the desired value of C:

$$C = \frac{1}{\ln\dfrac{4}{b-a}}\left[f(t_0) + \frac{1}{\pi^2}\int_a^b \ln\frac{1}{|t_0-t|}\,dt \int_a^b \frac{\sqrt{(b-\tau)(\tau-a)}\,f'(\tau)\,d\tau}{\sqrt{(b-t)(t-a)}\,(\tau-t)}\right]. \tag{1.6}$$

The right-hand side of (1.6) is a constant, which can be proved by differentiating this equality with respect to $t_0 \in [a,b]$ and applying the Poincaré–Bertrand formula that justifies the possibility of changing the order of integration in singular integrals [1].

Formula (1.5) for the solution to equation (1.1) is valid (and φ belongs to $\tilde{H}_\mu$ with some μ) when f' belongs to a wider class $\tilde{H}_{\mu,0}$ rather than to H_μ with the same μ,

Theorem 1 *Integral operator L is a continuous map of $\tilde{H}_\mu$ onto $\tilde{H}^1_{\mu,0}$ if $b-a \neq 4$ and $\mu < 1/2$.*

□ First, we will prove that L is an onto map. To this end, we must show that the solution to the equation

$$L\varphi = f$$

belongs to $\tilde{H}_\mu$ for any $f \in \tilde{H}^1_{\mu,0}$. As follows from representation (1.5), it is sufficient to prove that the integral

$$\Psi(t_0) = \frac{1}{\pi}\int_a^b \frac{\sqrt{(b-t)(t-a)}\varphi(t)dt}{t-t_0}$$

belongs to H_μ.

Since $\varphi \in \tilde{H}_{\mu,0}$, the function $\varphi^*(x) = R(x)\varphi(x)$ vanishes at the endpoints of segment $[a,b]$. Extend segment $[a,b]$ to a smooth closed curve $\hat{\Gamma}$ and continue function φ^* to $\hat{\Gamma}$ setting it equal to zero at $\hat{\Gamma} \setminus [a,b]$. Obviously, the continued function belongs to $H_\mu(\hat{\Gamma})$. According to Plemelj–Privalov theorem [1, 71], the singular integral operator with the Cauchy kernel is a continuous operator in the space $H_\mu(\hat{\Gamma})$ of functions satisfying the Hölder condition on the closed curve $\hat{\Gamma}$. Therefore, function $\Psi(t_0)$ belongs to $H_\mu(\hat{\Gamma})$ and, hence, to $H_\mu(\Gamma)$.

Now, let us prove that operator L is continuous from $\tilde{H}_\mu$ to $\tilde{H}^1_{\mu,0}$ (regardless of the value of $b-a$); i.e., for any $\varphi \in \tilde{H}_\mu$,

$$\|L\varphi\|_{\tilde{H}_\mu} \leq A\|\varphi\|_{\tilde{H}_\mu}, \tag{1.7}$$

where constant A does not depend on φ, and

$$\Psi(a) = \Psi(b) = 0.$$

The latter conditions follow from the estimates of the Cauchy integrals close to the endpoints of the curve of integration (see [1]). Thus, for any function φ that can be represented in the vicinity of endpoint a in the form

$$\varphi(t) = \frac{\varphi^*(t)}{(t-a)^\gamma}, \quad \gamma = \alpha + i\beta, \quad 0 < \alpha < 1, \tag{1.8}$$

the function

$$\Phi(t_0) = \frac{1}{\pi}\int_a^b \frac{\varphi(t)dt}{t-t_0},$$

admits, in the vicinity of a, the representation

$$\Phi(t_0) = \pm\frac{\cot\gamma\pi}{\pi}\frac{\varphi^*(a)}{(t_0-a)^\gamma} + \frac{\Phi^*(t_0)}{|t_0-a|^{\alpha_0}}, \tag{1.9}$$

where Φ^* satisfies the Hölder condition in the vicinity of a and $\alpha_0 < \alpha$.

According to the conditions of the theorem, function φ admits representation (1.9) with $\gamma = 1/2$ in the vicinity of points a and b; therefore, the first term in representation (1.9) vanishes, and function $R(t_0)\Phi(t_0)$ is equal to zero at a.

Now, let us prove estimate (1.7). Obviously,

$$\left|\frac{1}{\pi}\int_a^b \ln\frac{1}{|t_0-t|}\varphi(t)\,dt\right| \le \left|\frac{1}{\pi}\int_a^b \ln\frac{1}{|t_0-t|}\frac{dt}{\sqrt{(b-t)(t-a)}}\right| \|\varphi^*\|_C \le C_1\|\varphi^*\|_C \le C_1\|\varphi\|_{\tilde{H}_\mu}.$$

Therefore,

$$\|L\varphi\|_C \le C_1\|\varphi\|_{\tilde{H}_\mu}. \tag{1.10}$$

In order to prove that

$$\|S\varphi\|_{\tilde{H}_\mu} \le C_2\|\varphi\|_{\tilde{H}_\mu} = C_2\|\varphi^*\|_{H_\mu},$$

we represent function φ in the form (1.3) and consider operator S_1 defined by the relationship

$$(S_1\varphi^*)(t_0) \equiv R(t_0)(S\varphi)(t_0) = \sqrt{(b-t_0)(t_0-a)}\,\frac{1}{\pi}\int_a^b \frac{\varphi^*(t)dt}{\sqrt{(b-t)(t-a)}(t-t_0)}.$$

According to [73], operator S_1 is continuous in H_μ for any $\mu < 1/2$. Therefore,

$$\|S\varphi\|_{\tilde{H}_\mu} = \|RS\varphi\|_{H_\mu} \le A\|\varphi^*\|_{H_\mu} = A\|\varphi\|_{\tilde{H}_\mu} \tag{1.11}$$

with some $A > 0$

Inequalities (1.10) and (1.11) give the required estimate (1.7), which proves the continuity of L.

Since L is an onto operator for $b - a \ne 4$, the inverse operator L^{-1} is continuous. □

Theorem 2 *$L : \tilde{H}_\mu \to \tilde{H}^1_{\mu,0}$ is a Fredholm operator.*

□ For $b - a \ne 4$, L is a Fredholm operator because L is invertible. For $b - a = 4$, we represent L in the form

$$L\varphi \equiv \frac{1}{\pi}\int_a^b \left(\ln\frac{1}{|t_0-t|} + C_0\right)\varphi(t)dt - C_0\frac{1}{\pi}\int_a^b \varphi(t)dt \tag{1.12}$$

with a constant $C_0 \ne 0$. The first operator in (1.12) is invertible, which can be verified directly, and the second operator is one-dimensional. Therefore, L is a Fredholm operator, because it is represneted as a sum of invertible and one-dimensional operators. □

1.2 Integral operators with a logarithmic singularity in the Hölder spaces

Consider a more general form of the integral operator with a logarithmic singularity of the kernel

$$K\varphi \doteq L\varphi + N\varphi \equiv \int_a^b \left[\frac{1}{\pi}\ln\frac{1}{|t_0-t|} + N(t_0,t)\right]\varphi(t)dt,\ t_0 \in [a,b], \tag{1.13}$$

where $N(t_0,t)$ is a differentiable function whose derivative admits the representation

$$\frac{\partial N}{\partial t_0} = \frac{n(t_0,t)-n(t,t)}{t-t_0};\quad n(t_0,t) \in H_{\lambda,\lambda}([a,b]\times[a,b]),\ 0<\lambda<1. \tag{1.14}$$

The second condition in (1.14) means that, for the given function $n(t_0,t)$, there exist two positive constants A_1 and A_2 such that, for any points $x,y \in [a,b]$ and $t_1,t_2 \in [a,b]$,

$$|n(y,t_2)-n(x,t_1)| \le A_1|y-x|^\lambda + A_2|t_2-t_1|^\lambda. \tag{1.15}$$

Remark 1 *Sometimes, it is more convenient to use the equivalent representation of* $\frac{\partial N}{\partial t_0}$*:*

$$\frac{\partial N}{\partial t_0} = \frac{n(t_0,t)-n(t_0,t_0)}{t-t_0},\quad n(t_0,t) \in H_{\lambda,\lambda}([a,b]\times[a,b]),\ 0<\lambda<1.$$

Remark 2 *Instead of (1.14), we may assume that* $\frac{\partial N}{\partial t_0}$ *can be represented in the form*

$$\frac{\partial N}{\partial t_0} = \frac{n^*(t_0,t)}{|t-t_0|^\kappa},\quad 0\le\kappa<1, \tag{1.16}$$

with $n^*(t_0,t) \in H_{\lambda_1,\lambda_1}([a,b]\times[a,b]), 0<\lambda_1<1$.

Remark 3 *Most frequently, in representations (1.15),* $n(t,t) = n(t_0,t_0) = 0$.

Remark 4 *In many important cases,* $\partial N/\partial t_0$ *is a continuous function that belongs to* $H_{\lambda_2,\lambda_2}([a,b]\times[a,b])$, $0<\lambda_2<1$. *Then, in representation (1.14),*

$$n(t_0,t) = (t_0-t)n^*(t_0,t),\quad n^*(t_0,t) \in H_{\lambda_2,\lambda_2}([a,b]\times[a,b]),\ 0<\lambda_2<1.$$

Lemma 1 *Operator* $N: \tilde{H}_\mu \to \tilde{H}^1_{\mu,0}$ *is completely continuous.*

□ We will prove this lemma for the general case when representation (1.14) is valid. First, we show that operator N is continuous from $\tilde{H}_\mu$ to $\tilde{H}^1_\nu$ with some $\nu>\mu$.

It is sufficient to prove the estimate

$$\|N'\varphi\|_{H_\nu} \le A\|\varphi\|_{H_\mu} \tag{1.17}$$

for any $\varphi \in H_\mu$ with a constant $A>0$ independent of φ. Here,

$$N'\varphi = \int_a^b \frac{\sqrt{(b-t_0)(t_0-a)}}{\sqrt{(b-t)(t-a)}}\,\frac{n(t_0,t)-n(t,t)}{t-t_0}\,\varphi(t)\,dt$$

Using notation (1.4), we introduce the function

$$\Phi(t_0) = \int_a^b \frac{R(t_0)}{R(t)} \frac{n(t_0,t) - n(t,t)}{t - t_0} \varphi(t)\, dt.$$

The estimate

$$\max_{t_0 \in [a,b]} \Phi(t_0) \le C\|\varphi\|_C \le C\|\varphi\|_{H_\mu}$$

with C independent of φ can be easily obtained if we represent $\Phi(t_0)$ as the sum of integrals

$$\Phi(t_0) = \int_a^b \frac{(R(t_0) - R(t))(n(t_0,t) - n(t,t))}{R(t)(t - t_0)} \varphi(t)\, dt$$

$$+ \int_a^b \frac{n(t_0,t) - n(t,t)}{t - t_0} \varphi(t)\, dt = \Phi_1(t_0) + \Phi_2(t_0).$$

It is much more difficult to estimate $h_\nu(\Phi)$. We must prove that, for any $t_0, t_0+h \in \Gamma$,

$$|\Phi(t_0 + h) - \Phi(t_0)| \le C h^\nu,$$

where C is independent of t_0 and h.

First, we consider the difference $\Phi_1(t_0 + h) - \Phi_1(t_0)$:

$$\begin{aligned}\Phi_1(t_0 + h) - \Phi_1(t_0) = \int_a^b \frac{(R(t_0 + h) - R(t))(n(t_0 + h, t) - n(t,t))}{R(t)(t - t_0 - h)} \varphi(t)\, dt \\ - \int_a^b \frac{(R(t_0) - R(t))(n(t_0,t) - n(t,t))}{R(t)(t - t_0)} \varphi(t)\, dt.\end{aligned} \tag{1.18}$$

Divide the set of integration into two parts: the interval $l = (t_1, t_2)$ and the set $\Gamma' = (a,b) \setminus l$, where

$$t_1 = \max(a, t_0 - 2h), \quad t_2 = \min(b, t_0 + 2h). \tag{1.19}$$

Then, the difference $\Phi_1(t_0 + h) - \Phi_1(t_0)$ can be represented as

$$\Phi_1(t_0 + h) - \Phi_1(t_0) = I_0 + I_1,$$

where the expressions for I_0 and I_1, in which the integration is performed over l and Γ', respectively, are similar to (1.18).

Let us estimate I_0. Taking into account that $R \in H_{1/2}$ and $n(t_0,t) \in H_{\lambda,\lambda}([a,b] \times [a,b])$, we obtain

$$|I_0| \le \int_{t_1}^{t_2} \frac{|\varphi(t)|\, dt}{R(t)|t - t_0 - h|^{1/2-\lambda}} + \int_{t_1}^{t_2} \frac{|\varphi(t)|\, dt}{R(t)|t - t_0|^{1/2-\lambda}}.$$

Assume that $a < t_1 < t_2 \le (a+b)/2$ [1]. Then,

$$|I_0| \le \left(\frac{2}{b-a}\right)^{1/2} \|\varphi\|_C \int_{t_1}^{t_2} \frac{dt}{\sqrt{t-a}|t - t_0 - h|^{1/2-\lambda}} + \left(\frac{2}{b-a}\right)^{1/2} \int_{t_1}^{t_2} \frac{dt}{\sqrt{t-a}|t - t_0|^{1/2-\lambda}}$$

[1] The case $(a+b)/2 \le t_1 < t_2 < b$ is considered similarly. If $a < t_1(a+b)/2 < t_2 < b$, then we split each term in (1.18) into the integrals over $(t_1, (a+b))/2$ and $((a+b))/2, t_2)$ and estimate them seperately.

$$= \left(\frac{2}{b-a}\right)^{1/2} \|\varphi\|_C \int_{t_1}^{G_1} \frac{dt}{\sqrt{t-a}|t-t_0-h|^{1/2-\lambda}}$$

$$+ \left(\frac{2}{b-a}\right)^{1/2} \|\varphi\|_C \int_{G_1}^{t_2} \frac{dt}{\sqrt{t-a}|t-t_0-h|^{1/2-\lambda}}$$

$$+ \left(\frac{2}{b-a}\right)^{1/2} \|\varphi\|_C \int_{t_1}^{G_2} \frac{dt}{\sqrt{t-a}|t-t_0|^{1/2-\lambda}} + \left(\frac{2}{b-a}\right)^{1/2} \|\varphi\|_C \int_{G_2}^{t_2} \frac{dt}{\sqrt{t-a}|t-t_0|^{1/2-\lambda}},$$

where

$$G_1 = \max\left(t_1, \frac{a+t_0+h}{2}\right), \quad G_2 = \max\left(t_2, \frac{a+t_0}{2}\right).$$

If $G_1 = t_1$ or $G_2 = t_2$, then integrals over (t_1, G_1) or (t_2, G_2) vanish. As a result, we obtain

$$|I_0| \leq \left(\frac{2}{b-a}\right)^{1/2} \|\varphi\|_C \left(\int_{t_1}^{G_1} \frac{dt}{|t-a|^{1-\lambda}} + \int_{G_1}^{t_2} \frac{dt}{|t-t_0-h|^{1-\lambda}} \right.$$

$$\left. + \int_{t_1}^{G_2} \frac{dt}{|t-a|^{1-\lambda}} + \int_{G_2}^{t_2} \frac{dt}{|t-t_0|^{1-\lambda}} \right).$$

Each integral in brackets is estimated by $const \cdot h^\lambda$; for example,

$$\int_{t_1}^{G_1} \frac{dt}{|t-a|^{1-\lambda}} = \frac{1}{\lambda}\left((G_1-a)^\lambda - (t_1-a)^\lambda\right) \leq (G_1-t_1)^\lambda \leq (2h)^\lambda.$$

As a result, we obtain the estimate for I_0:

$$|I_0| \leq C_3 h^\lambda \|\varphi\|_C \leq C_3 h^\lambda \|\varphi\|_{H_\mu}. \tag{1.20}$$

Before estimating I_1, we represent it in the form

$$I_1 = \int_{\Gamma'} \left(\frac{(R(t_0+h) - R(t))}{R(t)(t-t_0-h)} - \frac{(R(t_0) - R(t))}{R(t)(t-t_0)} \right) (n(t_0+h,t) - n(t,t))\, \varphi(t)\, dt$$

$$+ \int_{\Gamma'} \frac{R(t_0) - R(t)}{R(t)(t-t_0)} (n(t_0+h,t) - n(t_0,t))\varphi(t)\, dt = I_2 + I_3.$$

First, we estimate integral I_3. Taking into account the Hölder properties of functions $R(t_0)$ and $n(t_0,t)$, we obtain

$$|I_3| \leq \int_{\Gamma'} \left| \frac{R(t_0) - R(t)}{R(t)(t-t_0)} \right| |(n(t_0+h,t) - n(t_0,t))|\, |\varphi(t)|\, dt \leq C_4 h^\lambda \|\varphi\|_C \int_{\Gamma'} \frac{dt}{R(t)|t-t_0|^{1/2}}$$

$$= C_4 h^\lambda \|\varphi\|_C \left(\int_a^{t_1} \frac{dt}{R(t)|t-t_0|^{1/2}} + \int_{t_2}^{b} \frac{dt}{R(t)|t-t_0|^{1/2}} \right) = C_4 h^\lambda \|\varphi\|_C (I_4 + I_5)$$

Assume that $a < t_1 < (a+b)/2$. Then,

$$|I_4| = \left| \int_a^{t_1} \frac{dt}{\sqrt{(b-t)(t-a)}|t-t_0|^{1/2}} \right| \le \left(\frac{2}{b-a}\right)^{1/2} \int_a^{t_1} \frac{dt}{(t-a)^{1/2}|t-t_0|^{1/2}}$$

$$= \left(\frac{2}{b-a}\right)^{1/2} \int_0^{t_1-a} \frac{dt}{t^{1/2}(t_0-a-t)^{1/2}}.$$

The change of varible

$$t = (t_0 - a)\rho, \tag{1.21}$$

in the last integral yields

$$\int_0^{t_1-a} \frac{dt}{t^{1/2}(t_0-a-t)^{1/2}} = \int_0^{\frac{t_1-a}{t_0-a}} \frac{d\rho}{\rho^{1/2}(1-\rho)^{1/2}} \le \int_0^1 \frac{d\rho}{\rho^{1/2}(1-\rho)^{1/2}} = C_5 = \text{const}.$$

Then,

$$|I_4| \le C_3 = \text{const}. \tag{1.22}$$

When estimating

$$I_5 = \int_{t_2}^{b} \frac{dt}{\sqrt{(b-t)(t-a)}|t-t_0|^{1/2}},$$

we must consider two cases: $t_2 < (a+b)/2$ and $(a+b)/2 \le t_2$.

In the first case $t_2 \ge (a+b)/2$, the estimate of integral I_5 is similar to that of integral I_4 for $t_1 \le (a+b)/2$. So, we will estimate I_5 only for $t_2 < (a+b)/2$ [note that this analysis is valid also for I_4 when $t_1 \le (a+b)/2$].

Represent I_5 as the sum

$$I_5 = \int_{t_2}^{(a+b)/2} \frac{dt}{\sqrt{(b-t)(t-a)}|t-t_0|^{1/2}} + \int_{(a+b)/2}^{b} \frac{dt}{\sqrt{(b-t)(t-a)}|t-t_0|^{1/2}}. \tag{1.23}$$

The second integral in (1.23) is estimated by analogy with integral I_4 for $t_1 \le (a+b)/2$. Therefore, we will estimate only the first integral:

$$\int_{t_2}^{(a+b)/2} \frac{dt}{\sqrt{(b-t)(t-a)}|t-t_0|^{1/2}} \le \left(\frac{2}{b-a}\right)^{1/2} \int_{t_2}^{(a+b)/2} \frac{dt}{(t-a)^{1/2}|t-t_0|^{1/2}}. \tag{1.24}$$

Performing the change of variable

$$t = t_0 + 2h\rho \tag{1.25}$$

in the last integral, we obtain

$$\begin{aligned} \int_{t_2}^{(a+b)/2} \frac{dt}{(t-a)^{1/2}|t-t_0|^{1/2}} &= \int_1^{((a+b)/2-t_0)/2h} \frac{d\rho}{\rho^{1/2}(\rho+(t_0-a)/2h)^{1/2}} \\ &\le \int_1^{((a+b)/2-t_0)/2h} \frac{d\rho}{\rho} = \ln\frac{(a+b)/2-t_0}{2h}. \end{aligned} \tag{1.26}$$

Since we assumed that $t_2 = t_0 + 2h < (a+b)/2$, the last term is estimated by $const \cdot |\ln h|$. Taking into account inequalities (1.22), (1.24), and (1.26), we obtain the following estimates for I_3:

$$|I_3| \leq C_6 \|\varphi\|_C \cdot h^\lambda |\ln h| \leq C_6 \|\varphi\|_C \cdot h^{\lambda-\varepsilon} \leq C_6 \|\varphi\|_{H_\mu} \cdot h^{\lambda-\varepsilon}, \qquad (1.27)$$

where ε is an arbitrarily small positive constant.

Now, let us estimate I_2. It can be represented as

$$I_2 = \int_{\Gamma'} \frac{(R(t_0+h) - R(t_0))}{R(t)(t-t_0-h)} (n(t_0+h,t) - n(t,t))\, \varphi(t)\, dt$$

$$+h \int_{\Gamma'} \frac{(R(t_0) - R(t))(n(t_0+h,t) - n(t,t))}{R(t)(t-t_0-h)(t-t_0)} \varphi(t)\, dt = I_6 + I_7.$$

Consider integral I_6:

$$|I_6| = \int_{\Gamma'} \frac{(R(t_0+h) - R(t_0))}{R(t)(t-t_0-h)} (n(t_0+h,t) - n(t,t))\, \varphi(t)\, dt$$

$$\leq h^{1/2} \|\varphi\|_C \int_{\Gamma'} \frac{dt}{R(t)|t-t_0-h|^{1-\lambda}}$$

$$= h^{1/2} \|\varphi\|_C \left(\int_a^{t_1} \frac{dt}{R(t)|t-t_0-h|^{1-\lambda}} + \int_{t_2}^{b} \frac{dt}{R(t)|t-t_0-h|^{1-\lambda}} = h^{1/2}\|\varphi\|_C (I_8 + I_9) \right).$$

If $\lambda \geq 1/2$, then the estimate of integrals I_8 and I_9 is similar to that of integrals I_4 and I_5. So we will consider only the case $\lambda < 1/2$. Assume that $t_1 \leq (a+b)/2$. Then,

$$I_8 = \int_a^{t_1} \frac{dt}{\sqrt{(b-t)(t-a)}|t-t_0-h|^{1-\lambda}} \leq \left(\frac{2}{b-a}\right)^{1/2} \int_a^{t_1} \frac{dt}{(t-a)^{1/2}|t-t_0-h|^{1-\lambda}}$$

$$= \left(\frac{2}{b-a}\right)^{1/2} \int_0^{t_1-a} \frac{dt}{t^{1/2}|t+a-t_0-h|^{1-\lambda}}.$$

Performing the change of variables (1.21) in the last integral, we obtain

$$\int_0^{t_1-a} \frac{dt}{t^{1/2}|t+a-t_0-h|^{1-\lambda}} = (t_0-a)^{\lambda-1/2} \int_0^{\frac{t_1-a}{t_0-a}} \frac{d\rho}{\rho^{1/2}|\rho - 1 - h/(t_0-a)|^{1-\lambda}}$$

$$\leq (t_0-a)^{\lambda-1/2} \int_0^1 \frac{d\rho}{\rho^{1/2}(\rho-1)^{1-\lambda}} \leq C_6 (t_0-a)^{\lambda-1/2}.$$

Assume that $a < t_1 = t_0 - 2h$ (otherwise, integral I_8 is absent). Then, $t_0 - a > 2h$, and the last integral is estimated by $C_6 \cdot (2h)^{\lambda-1/2}$ because $\lambda - 1/2 < 0$. Therefore,

$$I_8 \leq C_7 \cdot h^{\lambda-1/2}. \qquad (1.28)$$

Let us estimate I_9:

$$|I_9| = \int_{t_2}^{b} \frac{dt}{\sqrt{(b-t)(t-a)}|t-t_0|^{1-\lambda}}.$$

We must consider two cases: $t_2 < (a+b)/2$ and $(a+b)/2 \le t_2$.

In the case $t_2 \ge (a+b)/2$, integral I_9 is estimated like integral I_8 for $t_1 \le (a+b)/2$. So, we will estimate I_9 for $t_2 < (a+b)/2$ [and I_8 for $t_1 \le (a+b)/2$].

Represent I_9 as the sum

$$I_9 = \int_{t_2}^{(a+b)/2} \frac{dt}{\sqrt{(b-t)(t-a)}|t-t_0|^{1-\lambda}} + \int_{(a+b)/2}^{b} \frac{dt}{\sqrt{(b-t)(t-a)}|t-t_0|^{1-\lambda}}.$$

The last integral is estimated like integral I_8 for $t_1 \le (a+b)/2$; therefore, we will consider only the first integral:

$$\int_{t_2}^{(a+b)/2} \frac{dt}{\sqrt{(b-t)(t-a)}|t-t_0-h|^{1-\lambda}} \le \left(\frac{2}{b-a}\right)^{1/2} \int_{t_2}^{(a+b)/2} \frac{dt}{(t-a)^{1/2}|t-t_0-h|^{1-\lambda}}.$$

Performing the change of variable (1.25) in the last integral, we obtain

$$\begin{aligned}
&\int_{t_2}^{(a+b)/2} \frac{dt}{(t-a)^{1/2}|t-t_0-h|^{1-\lambda}} \\
&= (2h)^{\lambda-1/2} \int_{1}^{((a+b)/2-t_0)/2h} \frac{d\rho}{(\rho+(t_0-a)/2h)^{1/2}(\rho-1/2)^{1-\lambda}} \\
&\le (2h)^{\lambda-1/2} \int_{1}^{\infty} \frac{d\rho}{\rho^{1/2}(\rho-1/2)^{1-\lambda}} \le C_7 (2h)^{\lambda-1/2}.
\end{aligned} \tag{1.29}$$

The last inequality is valid because the last integral in (1.29) converges due to the condition $\lambda < 1/2$. Then, taking into account (1.28), we obtain

$$I_6 \le C_8 h^{\lambda} \|\varphi\|_C \le C_8 h^{\lambda} \|\varphi\|_{H_\mu} \tag{1.30}$$

Now, let us estimate integral I_7. We have the chain of inequalities

$$\begin{aligned}
|I_7| &\le h \int_{\Gamma'} \left| \frac{(R(t_0)-R(t))(n(t_0+h,t)-n(t,t))}{R(t)(t-t_0-h)(t-t_0)} \right| |\varphi(t)|\, dt \\
&\le \|\varphi(t)\|_C \cdot h \int_{\Gamma'} \frac{dt}{R(t)|t-t_0-h|^{1-\lambda}|t-t_0|^{1/2}}
\end{aligned}$$

$$= \|\varphi(t)\|_C \cdot h \left(\int_{a}^{t_1} \frac{dt}{R(t)|t-t_0-h|^{1-\lambda}|t-t_0|^{1/2}} + \int_{t_2}^{b} \frac{dt}{R(t)|t-t_0-h|^{1-\lambda}|t-t_0|^{1/2}} \right)$$

$$= \|\varphi(t)\|_C \cdot h(I_{10}+I_{11}).$$

Assume that $a < t_1 < (a+b)/2$ and consider integral I_{10}. Then,

$$I_{10} \leq \int_a^{t_1} \frac{dt}{R(t)|t-t_0-h|^{1-\lambda}|t-t_0|^{1/2}} \leq \left(\frac{2}{b-a}\right)^{1/2} \int_a^{t_1} \frac{dt}{(t-a)^{1/2}|t-t_0-h|^{1-\lambda}|t-t_0|^{1/2}}$$

$$= \left(\frac{2}{b-a}\right)^{1/2} \int_0^{t_1-a} \frac{dt}{t^{1/2}|t+a-t_0-h|^{1-\lambda}|t+a-t_0|^{1/2}}.$$

Change the variables in the last integral using formula (1.21):

$$\int_0^{t_1-a} \frac{dt}{t^{1/2}|t+a-t_0-h|^{1-\lambda}|t+a-t_0|^{1/2}}$$

$$= (t_0-a)^{\lambda-1} \int_0^{(t_1-a)/(t_0-a)} \frac{d\rho}{\rho^{1/2}|\rho-1-h/(t_0-a)|^{1-\lambda}|1-\rho|^{1/2}} = (t_0-a)^{\lambda-1} \cdot I_{12}$$

We assume that $t_0 - a > 2h$ (otherwise, $t_1 = a$ and integral I_{10} is absent) and consider two cases: $\dfrac{h}{t_0-a} > 1/4$ and $\dfrac{h}{t_0-a} \leq 1/4$. In the first case,

$$I_{12} \leq \int_0^{1/2} \frac{d\rho}{\rho^{1/2}(1-\rho)^{3/2-\lambda}} = C_8 = \text{const}.$$

In the second case,

$$I_{12} \leq \int_0^{1-\frac{2h}{t_0-a}} \frac{d\rho}{\rho^{1/2}(1-\rho)^{3/2-\lambda}} \leq \int_0^{1/2} \frac{d\rho}{\rho^{1/2}(1-\rho)^{3/2-\lambda}} + 2^{1/2} \int_{1/2}^{1-\frac{2h}{t_0-a}} \frac{d\rho}{(1-\rho)^{3/2-\lambda}}$$

$$= C_9 + 2^{1/2}\frac{1}{1/2-\lambda}(1-\rho)^{\lambda-1/2}\Bigg|_{1-\frac{2h}{t_0-a}}^{1/2} \leq C_{10}\left(\frac{2h}{t_0-a}\right)^{\lambda-1/2}.$$

Then, since $\dfrac{h}{t_0-a} < \dfrac{1}{2}$, we have

$$I_{12} \leq C_{10}2^{1/2-\lambda},$$

and

$$I_{10} \leq C_{11}\,(t_0-a)^{\lambda-1} < C_{11}(2h)^{\lambda-1}.$$

The last integral I_{11} is estimated in a similar manner. Thus, for I_7, we obtain the estimate

$$|I_7| \leq C_{12}\|\varphi\|_C \cdot h^{\lambda}. \tag{1.31}$$

As a result, using (1.20)–(1.31), one can estimate the difference $\Phi_1(t_0+h) - \Phi_1(t_0)$ as

$$|\Phi_1(t_0+h) - \Phi_1(t_0)| \leq C_{13}\|\varphi\|_C \cdot h^{\lambda-\varepsilon}, \tag{1.32}$$

where ε is an arbitrarily small constant.

Let us turn to estimating the difference $\Phi_2(t_0+h)-\Phi_2(t_0)$. It can be written as

$$\Phi_2(t_0+h)-\Phi_2(t_0)=\int_a^b\left(\frac{n(t_0+h,t)-n(t,t)}{t-t_0-h}-\frac{n(t_0,t)-n(t,t)}{t-t_0}\right)\varphi(t)\,dt. \tag{1.33}$$

Recalling the derivation of the estimate for $\Phi_1(t_0+h)-\Phi_1(t_0)$, we divide the set of integration into two parts: the interval $l=(t_1,t_2)$, where t_1 and t_2 are defined by formula (1.19), and the set $\Gamma'=(a,b)\setminus l$. Then, the difference $\Phi_2(t_0+h)-\Phi_2(t_0)$ can be represented as

$$\Phi_2(t_0+h)-\Phi_2(t_0)=J_0+J,$$

where the expressions for J_0 and J are similar to (1.33); the integration is performed over l and Γ', respectively.

First, we will estimate J_0. Taking into account that $n(t_0,t)\in H_{\lambda,\lambda}([a,b]\times[a,b])$, we obtain

$$|J_0|\le C_{14}\,\|\varphi\|_C\left(\int_l|t-t_0-h|^{\lambda-1}\,dt+\int_l|t-t_0|^{\lambda-1}\,dt\right)$$

$$\le C_{15}\,\|\varphi\|_C\frac{1}{\lambda}\left((t_2-t_0-h)^\lambda-(t_0+h-t_1)^\lambda+(t_2-t_0)^\lambda-(t_0-t_1)^\lambda\right)\le C_{16}h^\lambda\|\varphi\|_C.$$

In order to estimate J, we represent it as the sum

$$J=\int_{\Gamma'}\frac{n(t_0+h,t)-n(t_0,t)}{t-t_0}\varphi(t)\,dt$$

$$+\int_{\Gamma'}(n(t_0+h,t)-n(t,t))\left(\frac{1}{t-t_0-h}-\frac{1}{t-t_0}\right)\varphi(t)\,dt=J_1+J_2.$$

Integral J_1 is estimated as follows:

$$|J_1|\le C_{17}h^\lambda\|\varphi\|_C\left(\int_a^{t_1}\frac{dt}{|t-t_0|}+\int_{t_2}^b\frac{dt}{|t-t_0|}\right)=C_{17}h^\lambda\|\varphi\|_C\left(\ln\frac{|t_0-a|}{|t_0-t_1|}+\ln\frac{|b-t_0|}{|t_2-t_0|}\right). \tag{1.34}$$

If $t_0-a\le 2h$, the first logarithm in (1.34) vanishes due to the definition of t_1. If $t_0-a>2h$, the first logarithm is estimated by *const* $\cdot\,|\ln h|$. The same estimate holds for the second logarithm. Then,

$$|J_1|\le C_{18}h^\lambda\,|\ln h|\cdot\|\varphi\|_C\le C_{18}h^{\lambda-\varepsilon}\|\varphi\|_C\le C_{18}h^{\lambda-\varepsilon}\|\varphi\|_{H_\mu},$$

where ε is an arbitrarily small positive constant.

For J_2, we obtain

$$|J_2|=h\left|\int_{\Gamma'}\frac{n(t_0+h,t)-n(t,t)}{(t-t_0-h)(t-t_0)}\varphi(t)\,dt\right|$$

$$\le C_{19}h\|\varphi\|_C\int_{\Gamma'}\frac{dt}{|t-t_0-h|^{1-\lambda}|t-t_0|}\le C_{19}h\|\varphi\|_C\int_{\Gamma'}\frac{dt}{|t-t_0|^{2-\lambda}\left(1-\dfrac{h}{t-t_0}\right)^{1-\lambda}}.$$

Since $\dfrac{h}{|t-t_0|} \le \dfrac{1}{2}$ on Γ', we have

$$|J_2| \le C_{19} 2^{\lambda-1} h \|\varphi\|_C \int\limits_{\Gamma'} \frac{dt}{|t-t_0|^{2-\lambda}}$$
$$= 2^{\lambda-1} h \|\varphi\|_C \frac{1}{1-\lambda} \left((t_0-t_1)^{\lambda-1} - (t_0-a)^{\lambda-1} + (t_2-t_0)^{\lambda-1} - (b-t_0)^{\lambda-1} \right). \tag{1.35}$$

If $t_0 - a \le 2h$, the difference $(t_0-t_1)^{\lambda-1} - (t_0-a)^{\lambda-1}$ in (1.35) vanishes. If $t_0 - a > 2h$, this difference is estimated by *const* $\cdot\, h^{\lambda-1}$. The same estimate holds for the second logarithm. Then,

$$|J_2| \le C_{20} h^{\lambda} \|\varphi\|_C \le C_{20} h^{\lambda} \|\varphi\|_{H_\mu}.$$

As a result, the difference $\Phi_2(t_0+h) - \Phi_2(t_0)$ is estimated as

$$|\Phi_2(t_0+h) - \Phi_2(t_0)| \le C_{21} h^{\lambda-\varepsilon} \|\varphi\|_{H_\mu}, \tag{1.36}$$

where ε is an arbitrarily small constant.

Estimates (1.32) and (1.36) mean that

$$h_{\lambda-\varepsilon}(N'\varphi) \le C_{22} \|\varphi\|_{H_\mu}, \tag{1.37}$$

which yields inequality (1.17). Thus, operator N is continuous from H_μ to $\tilde{H}^1_{\lambda-\varepsilon}$, where ε is an arbitrarily small constant.

Now, choose the index $\nu = \lambda - \varepsilon$ greater than μ. Then, the operator $N : \tilde{H}_\mu \to \tilde{H}^1_\mu$ can be represented as a composition of the same operator from H_μ to $\tilde{H}_\nu$, which is continuous, and the operator of embedding of $\tilde{H}^1_\nu$ into $\tilde{H}^1_\mu$, which is completely continuous for $\nu > \mu$ according to [71]. Then, the operator $N : H_\mu \to \tilde{H}^1_\mu$ is completely continuous as a composition of continuous and completely continuous operators.

Finally, we will prove that N is a completely continuous operator from $\tilde{H}_\mu$ to $\tilde{H}^1_{\mu,0}$. Since $\tilde{H}^1_{\mu,0}$ is a closed subspace of $\tilde{H}^1_\mu$, it is sufficient to prove that the image of operator N defined on $\tilde{H}_\mu$ is a subspace of $\tilde{H}^1_{\mu,0}$.

Thus, it is sufficient to prove that, for any $\varphi \in \tilde{H}_\mu$, the order of singularity at points a and b is less than $1/2$.

With this purpose, we will estimate the integral

$$\Phi(t_0) = \int\limits_a^b \frac{n(t_0,t) - n(t,t)}{\sqrt{(b-t)(t-a)}(t-t_0)} \varphi(t)\, dt.$$

Taking into account the properties of functions $\varphi(t)$ and $n(t_0,t)$, we obtain

$$|\Phi(t_0)| \le C \|\varphi\|_C \int\limits_a^b \frac{dt}{\sqrt{(b-t)(t-a)}|t-t_0|^{1-\lambda}}.$$

Assume that $t_0 \le (a+b)/2$ (the opposite case is considered similarly). Divide the set of integration into three intervals: $(a, (a+t_0)/2)$, $((a+t_0)/2, (a+b)/2)$, and $((a+b)/2, b)$. Denote the coresponding integrals by K_1, K_2, and K_3.

First, let us estimate K_1:

$$K_1 = \int_a^{(a+t_0)/2} \frac{dt}{\sqrt{(b-t)(t-a)}|t-t_0|^{1-\lambda}} \le \left(\frac{2}{b-a}\right)^{1/2} \int_a^{(a+t_0)/2} \frac{dt}{\sqrt{t-a}|t-t_0|^{1-\lambda}}$$

$$\le \left(\frac{2}{b-a}\right)^{1/2} \left(\frac{2}{t_0-a}\right)^{1-\lambda} \int_a^{(a+t_0)/2} \frac{dt}{(t-a)^{1/2}} = 2\left(\frac{2}{b-a}\right)^{1/2} \left(\frac{2}{t_0-a}\right)^{1/2-\lambda}.$$

Hence, the order of singularity is less than 1/2.

Now, let us estimate K_2:

$$K_2 = \int_{(a+t_0)/2}^{(a+b)/2} \frac{dt}{\sqrt{(b-t)(t-a)}|t-t_0|^{1-\lambda}} \le \left(\frac{2}{b-a}\right)^{1/2} \int_{(a+t_0)/2}^{(a+b)/2} \frac{dt}{(t-a)^{1/2}|t-t_0|^{1-\lambda}}$$

$$= \left(\frac{2}{b-a}\right)^{1/2} \int_{(a+t_0)/2}^{(a+b)/2} \frac{dt}{(t-a)^{\varepsilon}(t-a)^{1/2-\varepsilon}|t-t_0|^{1-\lambda}},$$

where ε is an arbitrary constant, $0 < \varepsilon < \lambda$. Then,

$$K_2 \le \left(\frac{2}{b-a}\right)^{1/2} \left(\frac{2}{t_0-a}\right)^{1/2-\varepsilon} \int_{(a+t_0)/2}^{(a+b)/2} \frac{dt}{(t-a)^{\varepsilon}|t-t_0|^{1-\lambda}}$$

$$\le \left(\frac{2}{b-a}\right)^{1/2} \left(\frac{2}{t_0-a}\right)^{1/2-\varepsilon} \int_{(a+t_0)/2}^{(a+b)/2} \frac{dt}{|t-t_0|^{1+\varepsilon-\lambda}}$$

$$\le \frac{1}{\lambda-\varepsilon} \left(\frac{2}{b-a}\right)^{1/2} \left(\frac{2}{t_0-a}\right)^{1/2-\varepsilon} \left(\left(\frac{t_0-a}{2}\right)^{\lambda-\varepsilon} + \left(\frac{a+b}{2} - t_0\right)^{\lambda-\varepsilon}\right).$$

Hence, the singularity of K_2 at endpoint a is less than 1/2.

The estimate of integral K_3 is similar to that of K_1. As a result, we conclude that $\Phi(t_0) \in \tilde{H}^1_{\mu,0}$. □

Theorem 3 *$K : \tilde{H}_\mu \to \tilde{H}^1_{\mu,0}$ is a Fredholm operator.*

□ According to Theorem 2 and Lemma 1, operator K is the sum of the Fredholm and completely continuous operators. Therefore, it is a Fredholm operator. □

1.3 Logarithmic integral operators and Chebyshev polynomials

The Chebyshev polynomials have a deep connection with the logarithmic integral operators. In fact, if we separate the weight function, then it is easy to see that the Chebyshev polynomials of the first kind are the eigenfunctions of this logarithmic operator. Therefore, one can use the Chebyshev polynomials for analytical and numerical solution of integral equations with a logarithmic singularity of the kernel. We will prove the invertibility of the logarithmic operator in weighted Sobolev spaces and obtain the exact solution of the integral equation with a logarithmic singularity of the kernel in the form of the Fourier–Chebyshev series.

We note one more family of equations that can be considered in a similar manner; they are called hypersingular equations [4]. For these equations, we will obtain all the results listed above applying the same technique and using the Chebyshev polynomials of the second kind. In fact, for a hypersingular operator, the Chebyshev polynomials of the second kind are also the eigenfunctions. We consider both families in order to show a deep relation between these equations. Note that both these families may be considered from a single viewpoint as pseudodifferential equations [34].

Recall basic properties of the Chebyshev polynomials considered as the basis functions. The Chebyshev polynomials of the first kind, T_n, and the second kind, U_n, are defined by the standard formulas

$$T_0(x) = 1, \quad T_1(x) = x; \tag{1.38}$$

$$T_n(x) = 2xT_{n-1}(x) - T_{n-2}(x), \quad n = 2, 3, ...; \tag{1.39}$$

$$U_0(x) = 1, \quad U_1(x) = 2x; \tag{1.40}$$

$$U_n(x) = 2xU_{n-1}(x) - U_{n-2}(x), \quad n = 2, 3, \ldots . \tag{1.41}$$

If $|x| \leq 1$, then the following representations hold:

$$T_n(x) = \cos(n \arccos x), \quad n = 0, 1, ...; \tag{1.42}$$

$$U_n(x) = \frac{1}{\sqrt{1-x^2}} \sin((n+1) \arccos x), \quad n = 0, 1, \ldots . \tag{1.43}$$

For $x = \pm 1$, a transition to the limit should be performed in formula (1.43). The Chebyshev polynomials are coupled by the relationships

$$U_n(x) = \frac{1}{n+1} T'_{n+1}(x), \quad n = 0, 1, \ldots . \tag{1.44}$$

The Chebyshev polynomials are orthogonal functions with a weight on segment $[-1, 1]$:

$$\int_{-1}^{1} T_n(x) T_m(x) \frac{1}{\sqrt{1-x^2}} \, dx = \begin{cases} 0, & n \neq m \\ \pi/2, & n = m \neq 0 \\ \pi, & n = m = 0 \end{cases}; \tag{1.45}$$

$$\int_{-1}^{1} U_n(x) U_m(x) \sqrt{1-x^2} \, dx = \begin{cases} 0, & n \neq m \\ \pi/2, & n = m \end{cases}. \tag{1.46}$$

Thus, the Chebyshev polynomials of the first kind are orthogonal on $[-1, 1]$ with weight $1/\sqrt{1-x^2}$ and the Chebyshev polynomials of the second kind are orthogonal on $[-1, 1]$ with weight $\sqrt{1-x^2}$.

Formulas (1.45) and (1.46) may be used for expanding a function in the Fourier-Chebyshev series:

$$f(x) = \frac{a_0}{2} T_0(x) + \sum_{n=1}^{\infty} a_n T_n(x), \quad x \in [-1, 1]; \tag{1.47}$$

or

$$f(x) = \sum_{n=0}^{\infty} b_n U_n(x), \quad x \in [-1, 1]. \tag{1.48}$$

Coefficients a_n and b_n are defined by the formulas

$$a_n = \frac{2}{\pi}\int_{-1}^{1} f(x)T_n(x)\frac{dx}{\sqrt{1-x^2}}, \quad n = 0, 1, ...; \tag{1.49}$$

$$b_n = \frac{2}{\pi}\int_{-1}^{1} f(x)U_n(x)\sqrt{1-x^2}dx, \quad n = 0, 1, ... \tag{1.50}$$

Here, function $f(x)$ may take complex values; then coefficients a_n and b_n will be complex numbers.

In order to describe the convergence of the Fourier–Chebyshev series, we introduce the weighted spaces of functions

$$L_2^{(1)} = \{f(x) : \|f\|_1^2 := \int_{-1}^{1} |f(x)|^2 \frac{dx}{\sqrt{1-x^2}} < \infty\}$$

and

$$L_2^{(2)} = \{f(x) : \|f\|_2^2 := \int_{-1}^{1} |f(x)|^2 \sqrt{1-x^2}dx < \infty\}.$$

$L_2^{(1)}$ and $L_2^{(2)}$ are Hilbert spaces with the inner products

$$(f,g)_1 = \int_{-1}^{1} f(x)\overline{g(x)}\frac{dx}{\sqrt{1-x^2}}$$

and

$$(f,g)_2 = \int_{-1}^{1} f(x)\overline{g(x)}\sqrt{1-x^2}\,dx.$$

Thus, the Chebyshev polynomials specify orthogonal bases in spaces $L_2^{(1)}$ and $L_2^{(2)}$, respectively.

Define two weighted Sobolev spaces

$$\widetilde{W}_2^1 = \{f(x) : \ f \in L_2^{(1)}, \ f' \in L_2^{(2)}\}$$

and

$$\widehat{W}_2^1 = \{f(x) : \ f \in L_2^{(2)}, f^*(x) := f(x)\sqrt{1-x^2}, \ f^{*\prime} \in L_2^{(2)}, \ f^*(-1) = f^*(1) = 0\}.$$

They are also Hilbert spaces with the inner products

$$(f,g)_{\widetilde{W}_2^1} = \int_{-1}^{1} f(x)\frac{dx}{\sqrt{1-x^2}} \int_{-1}^{1} \overline{g(x)}\frac{dx}{\sqrt{1-x^2}} + \int_{-1}^{1} f'(x)\overline{g'(x)}\sqrt{1-x^2}\,dx$$

and

$$(f,g)_{\widehat{W}_2^1} = \int_{-1}^{1} (f(x)\sqrt{1-x^2})'\overline{(g(x)\sqrt{1-x^2})'}\sqrt{1-x^2}\,dx$$

and the norms

$$\|f\|^2_{\widetilde{W}^1_2} = \left| \int_{-1}^{1} f(x) \frac{dx}{\sqrt{1-x^2}} \right|^2 + \int_{-1}^{1} |f'(x)|^2 \sqrt{1-x^2}\, dx$$

and

$$\|f\|^2_{\widehat{W}^1_2} = \int_{-1}^{1} \left| (f(x)\sqrt{1-x^2})' \right|^2 \sqrt{1-x^2}\, dx.$$

In what follows, we will use two important formulas derived in [52]–[55]:

$$\int_{-1}^{1} \ln \frac{1}{|x-y|} T_n(x) \frac{dx}{\sqrt{1-x^2}} = \begin{cases} \pi \ln 2, & n = 0, \\ \pi T_n(y)/n, & n \geq 1; \end{cases} \quad -1 < y < 1 \tag{1.51}$$

and

$$\frac{d}{dy} \int_{-1}^{1} \frac{1}{x-y} U_n(x) \sqrt{1-x^2} dx = -\pi(n+1) U_n(y), \quad n \geq 0, \quad -1 < y < 1. \tag{1.52}$$

The integral on the left-hand side of formula (1.52) is understood in the sense of the principal value [1, 3].

Consider integral equations (1.1) and (1.13) with a logarithmic singularity of the kernel. Let us rewrite this equation in the form reduced to the standard interval of integration $(-1, 1)$ and with the explicitly separated weight function:

$$L\varphi = \int_{-1}^{1} \ln \frac{1}{|x-y|} \varphi(x) \frac{dx}{\sqrt{1-x^2}} = f(y), \quad -1 < y < 1, \tag{1.53}$$

and

$$(L+K)\varphi = \int_{-1}^{1} \left(\ln \frac{1}{|x-y|} + K(x,y)\right) \varphi(x) \frac{dx}{\sqrt{1-x^2}} = f(y), \quad -1 < y < 1. \tag{1.54}$$

Here, $K(x,y)$ is a known function without a singularity at $x = y$. In equations (1.53) and (1.54), we separate the weight function $1/\sqrt{1-x^2}$, which specifies the behaviour of the solution close to points -1 and $+1$. Operators L and $L+K$, which are defined by the left-hand side of equations (1.53) and (1.54), are considered as bounded operators in the appropriate spaces; namely, $L : L_2^{(1)} \to \widetilde{W}^1_2$ and $L + K : L_2^{(1)} \to \widetilde{W}^1_2$. Formula (1.51) means that the Chebyshev polynomials are eigenfunctions of logarithmic integral operator L.

If function $K(x,y)$ is sufficiently smooth (for example, continuously differentiable), then operator K is compact in these spaces. Consider the sufficient conditions that lead to the compactness of integral operator K.

Corollary 1 *Assume that*

$$K(x,y) \in L_2^{(1)} \times L_2^{(1)} \tag{1.55}$$

and

$$\frac{\partial K(x,y)}{\partial y} \in L_2^{(1)} \times L_2^{(2)}. \tag{1.56}$$

Then, the operator

$$K\varphi = \int_{-1}^{1} K(x,y)\varphi(x)\frac{dx}{\sqrt{1-x^2}},$$

$$K : L_2^{(1)} \to \widetilde{W}_2^1$$

is compact.

□ It follows from (1.55) that function $K(x,y)$ can be represented as the Fourier–Chebyshev series

$$K(x,y) = \sum_{n=0}^{\infty}\sum_{m=0}^{\infty} k_{nm}T_n(y)T_m(x).$$

Define the function

$$K_N(x,y) = \sum_{n=0}^{N}\sum_{m=0}^{\infty} k_{nm}T_n(y)T_m(x).$$

Denote $D_y \equiv \frac{\partial}{\partial y}$. It follows from (1.56) that

$$D_yK(x,y) = \sum_{n=1}^{\infty}\sum_{m=0}^{\infty} nk_{nm}U_{n-1}(y)T_m(x),$$

$$D_yK_N(x,y) = \sum_{n=1}^{N}\sum_{m=0}^{\infty} nk_{nm}U_{n-1}(y)T_m(x).$$

By virtue of (1.55), (1.56), and the Bessel inequality, we have

$$\sum_{n=0}^{\infty}\sum_{m=0}^{\infty} |k_{nm}|^2 < \infty, \tag{1.57}$$

$$\sum_{n=1}^{\infty}\sum_{m=0}^{\infty} n^2|k_{nm}|^2 < \infty. \tag{1.58}$$

Hence, for the remaining terms of the series, we obtain

$$\sum_{n=0}^{N}\sum_{m=0}^{\infty} |k_{nm}|^2 \to 0, \quad N\to\infty, \tag{1.59}$$

and

$$\sum_{n=1}^{N}\sum_{m=0}^{\infty} n^2|k_{nm}|^2 \to 0, \quad N\to\infty. \tag{1.60}$$

Let us define finite-dimensional operators K_N:

$$K_N\varphi = \int_{-1}^{1} K_N(x,y)\varphi(x)\frac{dx}{\sqrt{1-x^2}},$$

$$K_N : L_2^{(1)} \to \widetilde{W}_2^1.$$

Then, using the Schwartz inequality, we obtain

$$\begin{aligned}\|K\varphi - K_N\varphi\|^2_{\widetilde{W}_2^1} &= \|D_yK\varphi - D_yK_N\varphi\|_2^2 + |(K\varphi - K_N\varphi, 1)_1|^2 \\ &\le \Big(C_1\|D_yK(x,y) - D_yK_N(x,y)\|^2_{1,2} \\ &+ C_2\|K(x,y) - K_N(x,y)\|^2_{1,1}\Big)\,\|\varphi\|_1^2,\end{aligned} \tag{1.61}$$

where $\|\cdot\|_{i,j} \equiv \|\cdot\|_{L_2^{(i)}\times L_2^{(j)}}$ and constants C_1 and C_2 do not depend on φ and N. We also have the estimate

$$\|D_y K(x,y) - D_y K_N(x,y)\|_{1,2}^2 \le C_3 \sum_{n=1}^{N} \sum_{m=0}^{\infty} n^2 |k_{nm}|^2 \to 0, \quad N \to \infty,$$

and

$$\|K(x,y) - K_N(x,y)\|_{1,1}^2 \le C_4 \sum_{n=0}^{N} \sum_{m=0}^{\infty} |k_{nm}|^2 \to 0, \quad N \to \infty,$$

where constants C_3 and C_4 do not depend on N. Hence,

$$\|K - K_N\| \to 0, \quad N \to \infty.$$

Taking into account that K_N are finite-dimensional operators, we prove the compactness of K. □

Now, we will construct a solution to equation (1.53) in the form of series. Assume that function $f(y) \in \widetilde{W}_2^1$ and determine a solution $\varphi(x) \in L_2^{(1)}$ of equation (1.53) in the form

$$\varphi(x) = \frac{a_0}{2} T_0(x) + \sum_{n=1}^{\infty} a_n T_n(x), \tag{1.62}$$

where a_n are unknown coefficients. Expand function $f(y)$ in the Fourier–Chebyshev series

$$f(y) = \frac{f_0}{2} T_0(y) + \sum_{n=1}^{\infty} f_n T_n(y), \tag{1.63}$$

where f_n are given by formula (1.49). Substituting (1.62) and (1.63) into equation (1.53), we obtain the exact solution of equation (1.53) in the form of the series

$$\varphi(x) = \frac{f_0}{2\ln 2} T_0(x) + \sum_{n=1}^{\infty} n f_n T_n(x). \tag{1.64}$$

In some cases, the coefficients of series (1.64) can be calculated explicitly, and one can obtain an explicit solution of the problem.

Analysis of formula (1.64) shows that the following estimates hold for the solution of (1.53)

$$\|\varphi\|_{L_2^{(1)}} \le \|f\|_{\widetilde{W}_2^1} \le \sqrt{\pi}\ln 2 \|\varphi\|_{L_2^{(1)}}.$$

Thus, $L : L_2^{(1)} \to \widetilde{W}_2^1$ is a bounded operator and has a bounded inverse.

Theorem 4 *Operator $L : L_2^{(1)} \to \widetilde{W}_2^1$ is bounded and coutinuously invertible. The unique solution of equation (1.53) is given by formula (1.64).*

By analogy with logarithmic equations, we can consider singular equation (1.2) and the hypersingular equations [4]

$$H\varphi = \frac{d}{dy} \int_{-1}^{1} \frac{1}{x-y} \varphi(x) \sqrt{1-x^2}\, dx = f(y), \quad -1 < y < 1, \tag{1.65}$$

and

$$(H+K)\varphi = \frac{d}{dy}\int_{-1}^{1}\left(\frac{1}{x-y}+K(x,y)\right)\varphi(x)\sqrt{1-x^2}dx = f(y), \quad -1<y<1. \tag{1.66}$$

We do not differentiate under the integral sign because it yields a divergent integral. (1.65) and (1.66) are integrodifferential equations; they are usually called hypersingular equations. Operators H and $H+K$ are considered in the spaces $H : \widehat{W}_2^1 \to L_2^{(2)}$ and $H+K : \widehat{W}_2^1 \to L_2^{(2)}$. We will prove that these operators are bounded. If function $K(x,y)$ is sufficiently smooth (for example, continuously differentiable), then operator K is compact.

Formula (1.52) shows that the Chebyshev polynomials of the second kind are eigenfunctions of hypersingular operator H.

In order to solve equation (1.65), we will look for unknown function $\varphi(x)$ in the form

$$\varphi(x) = \sum_{n=0}^{\infty} b_n U_n(x). \tag{1.67}$$

Expand the right-hand side of the equation in the Fourier-Chebyshev series using the Chebyshev polynomials of the second kind,

$$f(x) = \sum_{n=0}^{\infty} f_n U_n(y), \tag{1.68}$$

where coefficients f_n are defined by formula (1.50).

Substituting these expressions into equation (1.65) and using formula (1.52), we obtain the exact solution

$$\varphi(x) = -\frac{1}{\pi}\sum_{n=0}^{\infty}\frac{1}{n+1}f_n U_n(x). \tag{1.69}$$

If function $f \in L_2^{(2)}$, then $\varphi \in \widehat{W}_2^1$, and vice versa. The solution of hypersingular equation (1.65) exists and is unique for any right-hand side $f \in L_2^{(2)}$. Moreover, from (1.69), we derive an equality for the norm of solution (1.65)

$$\|f\|_{L_2^{(2)}} = \frac{1}{\pi}\|\varphi\|_{\widehat{W}_2^1},$$

which proves that the operator $H : \widehat{W}_2^1 \to L_2^{(2)}$ is bounded and continuously invertible.

Theorem 5 *Operator $H : \widehat{W}_2^1 \to L_2^{(2)}$ is bounded and has a bounded inverse. The unique solution of equation (1.65) is given by formula (1.69).*

The above methods of solving the potential-type integral equations can be extended to other problems. In particular, the eigenvalue problems that are reduced to logarithmic or hypersingular integral equations can be solved using a similar technique. Such problems that arise in theoretical electromagnetics are considered in [34], [56]–[59], and in Chapter 4.

1.4 Integral operators with a logarithmic singularity of the kernel in the space of functions defined on a set of intervals

The results obtained above for integral operators (1.1) and (1.13) can be extended to the case when the set of integration consists of several intervals.

First, consider the equation

$$L\varphi \equiv \frac{1}{\pi}\int_{\Gamma} \ln\frac{1}{|t_0-t|}\varphi(t)dt = f(t_0), \quad t_0 \in \Gamma, \tag{1.70}$$

where $\Gamma = \bigcup_{i=1}^{n}\Gamma_i$, $\Gamma_i = [a_i, b_i]$, is a set of segments on the real axis. $\tilde{H}_\mu(\Gamma)$ and $\tilde{H}^1_{\mu,0}(\Gamma)$ are defined as the product spaces

$$\tilde{H}_\mu(\Gamma) = \prod_{i=1}^{n}\tilde{H}_\mu(\Gamma_i) \quad \text{and} \quad \tilde{H}^1_{\mu,0}(\Gamma) = \prod_{i=1}^{n}\tilde{H}^1_{\mu,0}(\Gamma_i)$$

Introducing the norms

$$\|\cdot\|_{\tilde{H}_\mu(\Gamma)} = \sum_{i=1}^{n}\|\cdot\|_{\tilde{H}_\mu(\Gamma_i)} \quad \text{and} \quad \|\cdot\|_{\tilde{H}^1_{\mu,0}(\Gamma)} = \sum_{i=1}^{n}\|\cdot\|_{\tilde{H}^1_\mu(\Gamma_i)},$$

we define Banach spaces $\tilde{H}_\mu(\Gamma)$ and $\tilde{H}^1_{\mu,0}(\Gamma)$.

Theorem 6 *Operator* L: $\tilde{H}_\mu(\Gamma) \longrightarrow \tilde{H}^1_{\mu,0}(\Gamma)$ *is continuous.*

□ Since $\tilde{H}_\mu(\Gamma)$ and $\tilde{H}^1_{\mu,0}(\Gamma)$ are product spaces, each element of these spaces can be treated as a vector. Thus, each $\varphi \in \tilde{H}_\mu(\Gamma)$ can be represented as

$$\varphi = (\varphi_1, \varphi_2, \ldots, \varphi_n)^T,$$

where $\varphi_i \in \tilde{H}_\mu(\Gamma_i)$ and $\varphi_i(x) \equiv \varphi(x)$, $x \in \Gamma_i$. Similarly, each $f \in \tilde{H}^1_{\mu,0}(\Gamma)$ can be represented as the vector

$$f = (f_1, f_2, \ldots, f_n)^T,$$

where $f_i \in \tilde{H}^1_{\mu,0}(\Gamma_i)$ and $f_i(t) \equiv f(t)$, $t \in \Gamma_i$.

Then, L can be treated as the matrix operator

$$L\varphi \equiv \begin{pmatrix} L_{11}\varphi_1 & + & L_{12}\varphi_2 & + & \ldots & + & L_{1n}\varphi_n \\ L_{21}\varphi_1 & + & L_{22}\varphi_2 & + & \ldots & + & L_{2n}\varphi_n \\ \vdots & & \vdots & & \ddots & & \vdots \\ L_{n1}\varphi_1 & + & L_{n2}\varphi_2 & + & \ldots & + & L_{nn}\varphi_n \end{pmatrix} = \begin{pmatrix} L_{11} & L_{12} & \ldots & L_{1n} \\ L_{21} & L_{22} & \ldots & L_{2n} \\ \vdots & \vdots & \ddots & \vdots \\ L_{n1} & L_{n2} & \ldots & L_{nn} \end{pmatrix}\begin{pmatrix} \varphi_1 \\ \varphi_2 \\ \vdots \\ \varphi_n \end{pmatrix},$$

where L_{ij} are the integral operators

$$L_{ij}\varphi \equiv \frac{1}{\pi}\int_{\Gamma_j} \ln\frac{1}{|t_0-t|}\varphi(t)dt, \quad t_0 \in \Gamma_i.$$

According to Theorem 1, each diagonal operator L_{ii}, $i = 1,\dots,n$, is a continuous operator from $\tilde{H}_\mu(\Gamma_i)$ to $\tilde{H}^1_{\mu,0}(\Gamma_i)$. Nondiagonal operators L_{ij}, $i \neq j$, are continuous from $\tilde{H}_\mu(\Gamma_j)$ to $\tilde{H}^1_{\mu,0}(\Gamma_i)$, because their kernels are infinitely smooth. Then, for any $\varphi \in \tilde{H}_\mu(\Gamma)$,

$$\|L\varphi\|_{\tilde{H}^1_\mu(\Gamma)} = \sum_{i=1}^{n} \|\sum_{j=1}^{n} L\varphi_j\|_{\tilde{H}^1_\mu(\Gamma_i)} \leq \sum_{i=1}^{n}\sum_{i=j}^{n} \|L\varphi_j\|_{\tilde{H}^1_\mu(\Gamma_i)} \leq C\sum_{i=1}^{n}\sum_{i=j}^{n} \|\varphi_j\|_{\tilde{H}_\mu(\Gamma_i)}$$

which proves the theorem. □

Now, let us study the solvability of equation (1.70). If f is a differentiable function, then any solution to (1.70) satisfies the singular integral equation

$$S\varphi \equiv \frac{1}{\pi}\int_\Gamma \frac{\varphi(t)dt}{t-t_0} = f'(t_0),\ t_0 \in \Gamma. \tag{1.71}$$

According to [1], if $f' \in H_\mu(\Gamma)$, then the general solution to (1.71) in $\tilde{H}_\mu(\Gamma)$ is given by the formula

$$\varphi(t) = L^{-1}f \equiv -\frac{1}{\pi}\int_\Gamma \frac{R(t_0)f'(t_0)dt_0}{R(t)(t_0-t)} + \frac{P_{n-1}(t)}{R(t)}, \tag{1.72}$$

where

$$R(t) = \prod_{i=1}^{n} \sqrt{(b_i - t)(t-a_i)}$$

and $P_n(t) = C_1 + C_2 t + \dots + C_n t^{n-1}$ is an arbitrary polynomial of order $n-1$ (the number of segments minus one).

Any solution to (1.70) is a solution to (1.71) (however, the inverse statement is not true). In order to satisfy equation (1.70), we substitute expression (1.72) into (1.70) to obtain

$$\sum_{j=1}^{n} C_j \frac{1}{\pi}\int_\Gamma \ln\frac{1}{|t_0-t|}\frac{t^{j-1}}{R(t)}dt = f(t_0) + \frac{1}{\pi^2}\int_\Gamma \ln\frac{1}{|t_0-t|}dt\int_\Gamma \frac{R(\tau)f'(\tau)d\tau}{R(t)(\tau-t)},\quad t_0 \in \Gamma. \tag{1.73}$$

The left- and right-hand sides of (1.73) are constant in each segment Γ_i. This statement can be easily verified by differentiating both sides of (1.73), applying the Poincaré–Bertrand formula of changing the order of integration in singular integrals [1], and taking into account that

$$\frac{1}{\pi}\int_\Gamma \frac{t^j dt}{R(t)(t-t_0)} = 0,\ t_0 \in \Gamma,\ j < n. \tag{1.74}$$

The latter equality can be proved using the residue theory. Therefore, choosing one point x_i at each interval (a_i, b_i), we obtain a system of n linear equations with respect to n unknowns—coefficients of polynomial P_{n-1}. The matrix of the system is formed by the elements

$$c_{ij} = \frac{1}{\pi}\int_\Gamma \ln\frac{1}{|x_i - t|}\frac{t^j}{R(t)}dt.$$

If the determinant of matrix $\|c_{ij}\|^n_{i,j=1}$ is not equal to zero (it is a property of the set of integration Γ), the solution to equation (1.70) exists and is unique for any $f \in H_{\mu,0}(\Gamma)$. The same is true for any $f \in \tilde{H}_{\mu,0}(\Gamma)$.

Since the determinant of $\|c_{ij}\|^n_{i,j=1}$ may vanish even when $n = 1$, the existence of inverse operator L^{-1} in the general case cannot be proved. However, L is always a Fredholm operator.

Theorem 7 *$L : \tilde{H}_\mu(\Gamma) \to \tilde{H}^1_{\mu,0}(\Gamma)$ is a Fredholm operator.*

□ Represent operator L as the sum $L = L_0 + L_C$, where L_0 is the diagonal operator

$$L\varphi = \begin{pmatrix} L_{11} & 0 & \dots & 0 \\ 0 & L_{22} & \dots & 0 \\ \vdots & \vdots & \ddots & \vdots \\ 0 & 0 & \dots & L_{nn} \end{pmatrix},$$

and $L_C = L - L_0$.

L_0 is a Fredholm operator. Indeed, if $b_i - a_i \neq 4$ for all i, operator L_0 is invertible, because all L_{ii} are invertible. If $b_i - a_i = 4$ for some i, then we can transform L_0 to a sum of invertible and finite-dimensional operators, representing the corresponding operator L_i by formula (1.12) with $C_0 \neq 1$.

Operator L_C, which can be represented in the form

$$L_C\varphi = \begin{pmatrix} 0 & L_{12} & \dots & L_{1n} \\ L_{21} & 0 & \dots & L_{2n} \\ \vdots & \vdots & \ddots & \vdots \\ L_{n1} & L_{n2} & \dots & 0 \end{pmatrix},$$

is completely continuous because all nondiagonal operators L_{ij}, $j \neq i$, have smooth kernels; therefore, they are completely continuous operators from $\tilde{H}_\mu(\Gamma_j)$ to $\tilde{H}^1_{\mu,0}(\Gamma_i)$. According to the Tikhonov theorem [74], the product of compact sets is compact in the product space. Hence, operator L_C maps any bounded set in $\tilde{H}_\mu(\Gamma_j)$ into a compact set in $\tilde{H}^1_{\mu,0}(\Gamma_i)$, which means that L_C is a compact (completely continuos) operator. Then, L is a Fredholm operator, because it is represented as a sum of Fredholm and completely continuos operators. □

Now, consider the integral operator

$$K\varphi = L\varphi + N\varphi \equiv \int_\Gamma \left[\frac{1}{\pi}\ln\frac{1}{|t_0 - t|} + N(t_0, t)\right]\varphi(t)dt, \quad t_0 \in \Gamma,$$

where $N(t_0, t)$ is represented in the form (1.14) with $n(t_0, t) \in H_{\lambda\times\lambda}(\Gamma \times \Gamma)$.

Operator K can be written in the matrix form

$$K\varphi = \begin{pmatrix} K_{11} & K_{12} & \dots & K_{1n} \\ K_{21} & K_{22} & \dots & K_{2n} \\ \vdots & \vdots & \ddots & \vdots \\ K_{n1} & K_{n2} & \dots & K_{nn} \end{pmatrix},$$

where

$$K_{ij}\varphi = L_{ij}\varphi + N_{ij}\varphi \equiv \frac{1}{\pi}\int_{\Gamma_j} \left[\ln\frac{1}{|t_0 - t|} + N(t_0, t)\right]\varphi(t)dt, \quad t_0 \in \Gamma,$$

Lemma 2 *Operator $N : \tilde{H}_\mu(\Gamma) \to \tilde{H}^1_{\mu,0}(\Gamma)$ is completely continuous.*

□ Represent N in the matrix form. According to Lemma 1, each operator $N_{ij} : \tilde{H}_\mu(\Gamma_j) \to \tilde{H}^1_{\mu,0}(\Gamma_i)$ is completely continuos. Literally repeating the proof of Theorem 7, we conclude that operator N is completely continuous in the product space. □

Theorem 8 *$K : \tilde{H}_\mu(\Gamma) \to \tilde{H}^1_{\mu,0}(\Gamma)$ is a Fredholm operator.*

□ Represent operator K as the sum

$$K = L_0 + L_C + N.$$

The diagonal operator L_0 is a Fredholm operator, and the sum $L_C + N$ is a completely continuos operator. Hence, K is a Fredholm operator. □

1.5 Integral operators with fixed logarithmic singularities of the kernel

Some problems of mathematical physics are reduced to integral equations, in which the integral operator has additional logarithmic singularities of the kernel. For example, the problem of diffraction by the open end of a waveguide with a flange can be reduced to the equation

$$K\varphi \equiv -\frac{1}{\pi}\int_0^\pi \left((1+\varepsilon)\ln\left|\sin\frac{x_0-x}{2}\right| + \varepsilon\ln\left|\sin\frac{x_0+x}{2}\right| + N(x_0,x)\right)\varphi(x)dx = f(x_0),$$
$$x_0 \in [0,\pi], \quad \varepsilon \geq 1 \tag{1.75}$$

where and $N(s, s_0)$ is a differentiable function satisfying conditions (1.14) and (1.15).

In addition to the logarithmic singularity at $x = x_0$, the kernel of operator K has fixed (immovable) logarithmic singularities at $x = x_0 = 0$ and $x = x_0 = \pi$. In fact, operator K can be represented in the form

$$K\varphi \equiv \frac{1}{\pi}\int_0^\pi \left((1+\varepsilon)\ln\frac{1}{|x_0-x|} + \varepsilon\ln\frac{1}{x_0+x} + \varepsilon\ln\frac{1}{2\pi - x_0 - x} + N(x_0,x)\right)\varphi(x)dx$$

with another function $N(s, s_0)$ satisfying (1.14) and (1.15).

The presence of fixed logarithmic singularities strongly influences the properties of the corresponding integral operators. First of all, the class of solvability of equation (1.75) differs from that of equation (1.1). Let us consider operator L, the principal part of K:

$$L\varphi = -\frac{1}{\pi}\int_0^\pi \left((1+\varepsilon)\ln\left|\sin\frac{x_0-x}{2}\right| + \varepsilon\ln\left|\sin\frac{x_0+x}{2}\right|\right)\varphi(x)dx$$

Denote by $\tilde{H}_{\delta,\alpha}$ the set of functions φ that admit the representation

$$\varphi(x) = \frac{\varphi^*(x)}{x^{-\alpha}(\pi - x)^{-\alpha}},$$

where $\varphi^* \in H_\delta([0,\pi])$. The set $\tilde{H}_{\delta,\alpha}$ is a Banach space with the norm

$$\|\varphi\|_{\tilde{H}_{\delta,\alpha}} = \|\varphi^*\|_{H_\delta}.$$

Denote by $\tilde{H}^1_{\delta,\alpha}$ the set of functions, whose derivatives belong to $\tilde{H}_{\delta,\alpha}$. The norm in this space is defined as

$$\|\varphi\|_{\tilde{H}_{\delta,\alpha}} = \|\varphi\|_C + \|\varphi'\|_{\tilde{H}^1_{\delta,\alpha}}.$$

Theorem 9 ([75]) *The integral operator L is a continuous map of $\tilde{H}_{\delta,\alpha}$ onto $\tilde{H}^1_{\delta,\alpha}$ for any $\delta < \alpha$, where*

$$\alpha = \frac{1}{\pi}\arctan\sqrt{1+2\varepsilon}.$$

The inverse operator L^{-1} is defined by the formulas

$$(L^{-1}f)(x) = -\frac{1}{2\pi(1+\varepsilon)}\left(\cot\frac{x}{2}\right)^{1-2\alpha}\int_0^\pi\left(\tan\frac{x_0}{2}\right)^{1-2\alpha}\frac{\sin x_0\, f'(x_0)\, dx_0}{\cos x_0 - \cos x}$$

$$-\frac{1}{2\pi(1+\varepsilon)}\left(\tan\frac{x}{2}\right)^{1-2\alpha}\int_0^\pi\left(\cot\frac{x_0}{2}\right)^{1-2\alpha}\frac{\sin x_0\, f'(x_0)\, dx_0}{\cos x_0 - \cos x}$$

$$-C\left(\left(\tan\frac{x}{2}\right)^{1-2\alpha} + \left(\cot\frac{x}{2}\right)^{1-2\alpha}\right),$$

where

$$C = -\frac{1+2\varepsilon}{2\pi^2(1+\varepsilon)C_0}\left\{\int_0^\pi \ln\left|\sin\frac{x}{2}\right|\left(\left(\cot\frac{x}{2}\right)^{1-2\alpha}\int_0^\pi\left(\tan\frac{\sigma}{2}\right)^{1-2\alpha}\frac{\sin\sigma\, f'(\sigma)\, d\sigma}{\cos\sigma - \cos x}\right.\right.$$

$$\left.\left. + \left(\tan\frac{x}{2}\right)^{1-2\alpha}\int_0^\pi\left(\cot\frac{\sigma}{2}\right)^{1-2\alpha}\frac{\sin\sigma\, f'(\sigma)\, d\sigma}{\cos\sigma - \cos x}\right) dx + 4\pi(1+\varepsilon)f(0)\right\},$$

$$C_0 = -\frac{1+2\varepsilon}{\pi}\int_0^\pi \ln\left|\sin\frac{x}{2}\right|\left(\left(\tan\frac{x}{2}\right)^{1-2\alpha} + \left(\cot\frac{x}{2}\right)^{1-2\alpha}\right) dx.$$

□ Consider the equation

$$-\frac{1}{\pi}\int_0^\pi\left((1+\varepsilon)\ln\left|\sin\frac{x_0 - x}{2}\right| + \varepsilon\ln\left|\sin\frac{x_0+x}{2}\right|\right)\varphi(x)dx = f(x_0),\ x_0 \in [0,\pi]. \quad (1.76)$$

Differentiate it with respect to x_0:

$$-\frac{1}{\pi}\int_0^\pi\left((1+\varepsilon)\cot\frac{x_0-x}{2} + \varepsilon\cot\frac{x_0+x}{2}\right)\varphi(x)dx = f'(x_0). \quad (1.77)$$

Following [76], we introduce the new variables

$$t_0 = \cos x_0, \quad t = \cos x$$

and the new unknown function

$$F(t) = \frac{1}{2}\varphi(x)\frac{dx}{dt},$$

to obtain the equation

$$\frac{1}{\pi}\int_{-1}^1\frac{F(t)}{t-t_0}\left((1-t^2)^{1/2} + (1+2\varepsilon)(1-t_0^2)^{1/2}\right)dt = -f'(x_0(t_0)). \quad (1.78)$$

In some special cases, equation (1.78) has been solved explicitly [76]. Here, we use another approach [75] and reduce the equation to the Riemann vector problem [2].

Write equation (1.78) in the equivalent form

$$\frac{\tau}{\pi i}\int_{-1}^{1}\frac{F(t)\,dt}{t-t_0}+\frac{1}{\pi i\sqrt{1-t_0^2}}\int_{-1}^{1}\frac{\sqrt{1-t^2}F(t)\,dt}{t-t_0}=g(t_0), \tag{1.79}$$

where $i=\sqrt{-1}$, $\tau=1+2\varepsilon$, and

$$g(t_0)=\frac{i}{\sqrt{1-t_0^2}}\frac{df}{dx_0}(x_0(t_0)).$$

Introduce two analytical functions of complex variable z

$$\begin{aligned}\Phi_1(z)&=\frac{1}{2\pi i}\int_{-1}^{1}\frac{F(t)\,dt}{t-z},\\ \Phi_2(z)&=\frac{1}{2\pi i\sqrt{1-z^2}}\int_{-1}^{1}\frac{\sqrt{1-t^2}\,F(t)\,dt}{t-z}.\end{aligned} \tag{1.80}$$

The branch of the square root is chosen so that its limit on the upper side of the cut along the segment $[-1,1]$ is nonnegative. According to Plemelj–Sokhotski theorem [1], the limit values of functions Φ_1 and Φ_2 for $t_0\in(-1,1)$ are coupled by the relationships

$$\begin{aligned}&\Phi_1^+(t_0)+\Phi_1^-(t_0)=\frac{1}{\pi i}\int_{-1}^{1}\frac{F(t)\,dt}{t-t_0},\\ &\Phi_1^+(t_0)-\Phi_1^-(t_0)=F(t_0),\\ &\Phi_2^+(t_0)+\Phi_2^-(t_0)=F(t_0),\\ &\Phi_2^+(t_0)-\Phi_2^-(t_0)=\frac{1}{\pi i\sqrt{1-t_0^2}}\int_{-1}^{1}\frac{\sqrt{1-t^2}F(t)\,dt}{t-t_0},\end{aligned}$$

where indices + and − denote the limit values on the upper and lower sides of the interval $(-1,1)$, respectively.

Then equation (1.79) can be written as

$$\tau\left(\Phi_1^+(t_0)-\Phi_1^-(t_0)\right)+\left(\Phi_2^+(t_0)-\Phi_2^-(t_0)\right)=g(t_0).$$

Taking into account that

$$\Phi_1^+(t_0)-\Phi_1^-(t_0)=\Phi_2^+(t_0)+\Phi_2^-(t_0),$$

we arrive at the following Riemann problem: find two analytical functions $\Phi_1(z)$ and $\Phi_2(z)$ vanishing at infinity and satisfying the boundary conditions on the curve $\Gamma=(-1,1)$

$$\begin{pmatrix}\Phi_1^+\\ \Phi_2^+\end{pmatrix}=G\begin{pmatrix}\Phi_1^-\\ \Phi_2^-\end{pmatrix}+\frac{1}{1+\tau}\begin{pmatrix}g\\ g\end{pmatrix}, \tag{1.81}$$

where

$$G = \begin{pmatrix} 1-\tau & 2 \\ -2\tau & 1-\tau \end{pmatrix}. \tag{1.82}$$

First, let us consider the uniform problem

$$\begin{pmatrix} X_1^+ \\ X_2^+ \end{pmatrix} = G \begin{pmatrix} X_1^- \\ X_2^- \end{pmatrix}. \tag{1.83}$$

The complete set of solutions to problem (1.83) is called the canonical set of solutions to the vector Riemann problem [2]. Any solution to nonuniform problem (1.81) can be expressed by means of the canonical system and right-hand side g.

In the general case, according to [1, 2], the canonical system cannot be found in a closed form. However, for the given matrix G specified by (1.82), we can construct the canonical system.

Let us determine the solutions to uniform problem (1.83) in the class of functions unbounded in the vicinity of endpoints -1 and 1 and integrable in the interval $(-1,1)$. Assume that one of such solutions has the form

$$X(z) = (1-z)^{-\nu}(1+z)^{\nu-1} \begin{pmatrix} a \\ b \end{pmatrix},$$

where a, b, and $0 < \nu < 1$ are undetermined constants. Then, vector-valued function $X(z)$ is holomorphic on the complex plane cut along the segment $[-1,1]$, has the order $O(z^{-1})$ at infinity, and is integrable on the interval $(-1,1)$. We choose the branch of the function $(1-z)^{-\nu}(1+z)^{\nu-1}$, which equals $|1-z|^{-\nu}|1+z|^{\nu-1}$ on the upper side of the cut along $(-1,1)$. The limit values of the function X on the upper and lower sides of $(-1,1)$ are

$$X^+(t) = |1-z|^{-\nu}|1+z|^{\nu-1} \begin{pmatrix} a \\ b \end{pmatrix}$$

and

$$X^-(t) = e^{2i\pi\nu}|1-z|^{-\nu}|1+z|^{\nu-1} \begin{pmatrix} a \\ b \end{pmatrix}.$$

Since $X^+(t)$ and $X^-(t)$ are coupled by matrix G, the problem is reduced to the determination of a, b, and ν such that $e^{2i\pi\nu}$ would be an eigenvalue of matrix G and $(a,b)^T$ would be a corresponding eigenvector. In other words, the problem is reduced to the equation

$$G \begin{pmatrix} a \\ b \end{pmatrix} = e^{2i\pi\nu} \begin{pmatrix} a \\ b \end{pmatrix}$$

with respect to unknowns a, b, and ν.

Represent τ in the form

$$\tau = (\tan \pi\mu)^2.$$

Then, we can find two eigenvalues of matrix G:

$$\lambda_{1,2} = e^{\pm 2i\pi\mu},$$

which gives one possible value of ν,

$$\nu = \mu = \arctan\sqrt{\tau}$$

The corresponding eigenvector is

$$\begin{pmatrix} a \\ b \end{pmatrix} = \begin{pmatrix} \cos \pi\mu \\ -i \sin \pi\mu \end{pmatrix}$$

Thus, we have found one of the solutions to uniform problem (1.83). The second solution is determined in a similar manner. As a result, we obtain the following canonical matrix [the matrix of solutions to uniform problem (1.83)]:

$$\mathrm{X}(z) = \begin{pmatrix} \cos \pi\mu (1-z)^{\mu-1}(1+z)^{-\mu} & \cos \pi\mu (1-z)^{-\mu}(1+z)^{1-\mu} \\ i \sin \pi\mu (1-z)^{\mu-1}(1+z)^{-\mu} & -i \sin \pi\mu (1-z)^{-\mu}(1+z)^{1-\mu} \end{pmatrix}.$$

Evidently, the determinant of matrix $\mathrm{X}(z)$ is not equal to zero in the whole complex plane and

$$\lim_{z\to\infty} z^2 \mathrm{X}(z) = -i \sin \pi\mu \neq 0,$$

because $\mu < 1/2$. Hence, in accordance with the definition [1, 2], $\mathrm{X}(z)$ is a canonical matrix The general solution to problem (1.81) can be represented in the form

$$\Phi(z) = \frac{\mathrm{X}(z)}{2\pi i} \int_{-1}^{1} [\mathrm{X}^+(t)]^{-1} \tilde{g}(t) \frac{dt}{t-z} + \mathrm{X}(z) P(z),$$

where both components of vector $\tilde{g}(t)$ are equal to $g(t)/(1+\tau)$, and $P(z)$ is an undetermined polynomial vector. The choice of the latter must ensure the required order of $\Phi_1(z)$ and $\Phi_2(z)$ at infinity. As follows from representations (1.80),

$$\Phi_1(z) = O(z^{-1}) \quad \text{and} \quad \Phi_2(z) = O(z^{-2}).$$

Satisfying these conditions, we find that

$$P(z) = \begin{pmatrix} Ce^{i\pi\mu} \\ -Ce^{-i\pi\mu} \end{pmatrix},$$

where C is an arbitrary constant.

Using the Sokhotskii formulas [1, 2], after some algebra, we obtain

$$\begin{aligned} F(t_0) &= \Phi_1^+(t_0) - \Phi_1^-(t_0) = \\ &= \frac{1}{2\pi i(1+\tau)(1-t_0)^{1-\mu}(1+t_0)^{\mu}} \int_{-1}^{1} \frac{(1-t)^{1-\mu}(1+t)^{\mu} g(t)\, dt}{t-t_0} + \\ &+ \frac{1}{2\pi i(1+\tau)(1-t_0)^{\mu}(1+t_0)^{1-\mu}} \int_{-1}^{1} \frac{(1-t)^{\mu}(1+t)^{1-\mu} g(t)\, dt}{t-t_0} + \\ &+ C\left(\frac{1}{(1-t_0)^{1-\mu}(1+t_0)^{\mu}} + \frac{1}{(1-t_0)^{\mu}(1+t_0)^{1-\mu}} \right). \end{aligned} \qquad (1.84)$$

Formula (1.36) gives the general solution to equation (1.81). Taking into account that

$$g(t) = i f'(x)(1-x^2)^{-1/2},$$
$$\varphi(x) = -2F(t)(1-t^2)^{1/2}$$

and returning to variables x, x_0, we obtain the general solution to (1.77) in the form

$$\varphi(x) = -\frac{1}{2\pi(1+\varepsilon)}\left(\cot\frac{x}{2}\right)^{1-2\mu}\int_0^\pi\left(\tan\frac{x_0}{2}\right)^{1-2\mu}\frac{\sin x_0\, f'(x_0)\, dx_0}{\cos x_0 - \cos x}$$
$$-\frac{1}{2\pi(1+\varepsilon)}\left(\tan\frac{x}{2}\right)^{1-2\mu}\int_0^\pi\left(\cot\frac{x_0}{2}\right)^{1-2\mu}\frac{\sin x_0\, f'(x_0)\, dx_0}{\cos x_0 - \cos x} \tag{1.85}$$
$$-C\left(\left(\tan\frac{x}{2}\right)^{1-2\mu} + \left(\cot\frac{x}{2}\right)^{1-2\mu}\right).$$

Now, we substitute formula (1.85) into equation (1.77) and determine constant C so that φ would satisfy (1.76). For any right-hand side f, this constant is determined uniquely, because

$$C_0 = -\frac{1+2\varepsilon}{\pi}\int_0^\pi\left((1+\varepsilon)\ln\left|\sin\frac{x_0-x}{2}\right| + \varepsilon\ln\left|\sin\frac{x_0+x}{2}\right|\right)$$
$$\left(\left(\tan\frac{x}{2}\right)^{1-2\alpha} + \left(\cot\frac{x}{2}\right)^{1-2\alpha}\right)dx, \tag{1.86}$$

is a nonzero constant. Indeed, $C_0 > 0$, because the integrand in (1.86) is negative for $x_0 \in [0, \pi]$. The equality $\dfrac{dC_0}{dx_0} = 0$ follows from the procedure of solution.

Thus, we have proved that inverse operator L^{-1} exists. The continuity of operator $L : \tilde{H}_{\mu,\alpha} \to \tilde{H}^1_{\mu,\alpha}$ for $\mu < \alpha$ can be proved by analogy with Theorem 1. □

Now let us turn to operator K specified by formula (1.75).

Theorem 10 *$K : \tilde{H}_{\mu,\alpha} \to \tilde{H}^1_{\mu,\alpha}$, $\mu < \alpha$, is a Fredholm operator.*

□ The complete continuity of operator N is proved by analogy with Lemma 1. Then, K is a Fredholm operator because it is represented as a sum of Fredholm and completely continuous operators. □

1.6 Logarithmic integral operator-valued functions: elements of the spectral theory

We will formulate basic definitions and notions of the spectral theory of operators that will be used in the subsequent analysis. For details, one may refer to any textbook on functional analysis (for example, [50] or [74]) or original papers and monographs ([26], [35], or [37]).

Let G be an open connected set of complex plane C. Operator $A(z)$, $z \in G$, is called an operator-valued function on Banach space X, if for each $z \in G$, $A(z)$ is a linear bounded operator acting from Banach space X to Banach space Y with the domain $D(A) \subseteq X$ and range $R(A) \subseteq Y$.

Definition 1 *Operator-valued function $A(z)$ is called holomorphic at a point $z_0 \in G$ if for a certain $p > 0$, $A(z)$ can be expanded into the series*

$$A(z) = A(z_0) + \sum_{m=1}^{\infty}(z - z_0)^m A_m, \tag{1.87}$$

which converges in the operator norm. The convergence of series (1.89) for $|z - z_0| < p$ *is equivalent to the fact that for every* $p_1 > 0$ *and* $p_1 < p$*, there exists a positive constant* M *such that* $||A_m|| < Mp_1^{-m}$*,* $m = 1, 2, \ldots$ *A vector-valued function* $\varphi(z)$*—an element of Banach space* X*—is defined in a similar manner.*

Definition 2 z_0 *is called the characteristic number (CN) of operator-valued function* $A(z)$ *if* $A(z)$ *is holomorphic in a perforated vicinity of* z_0 $0 < |z - z_0| < p$ *for a certain* $p > 0$*, and there exists a vector-valued function* $\varphi(z)$ *(an element of Banach space* X*), which is holomorphic at point* z_0*, and vector-valued function* $A(z)\varphi(z)$ *(an element of Banach space* Y*) is also holomorphic at* z_0 *and vanishes at this point,*

$$A(z_0)\varphi(z_0) = 0 \tag{1.88}$$

(here, 0 is the zero element of Y*).* $\varphi(z)$ *is called the root function of* $A(z)$ *corresponding to* z_0*.*

Definition 3 *The order of* z_0 *considered as a zero of vector-valued function* $A(z)\varphi(z)$ *is called the multiplicity of the root function* $\varphi(z)$*, and* $\varphi_0 = \varphi(z_0)$ *is called the eigenvector of* $A(z)$*.*

The set of all solutions to equation (1.88) forms a linear manifold Z_A.

Let X^* be the conjugate space to X (i.e., the set of all functionals on X) and let $z = z_0$, where z_0 is a CN of operator-valued function A. Denote by Z_A^* the linear manifold of the elements from X^* that are orthogonal to $R(A)$ [i.e., $Z_A^* = \{f \in X^* : f(\psi) = 0 \ \forall \psi \in R(A)\}$]. Then, according to the standard terminology, $\ker A = \overline{Z_A}$ is the kernel of A, and $\overline{Z_A^*} = Y \backslash \overline{R(A)}$ is the defect space of A.

Set

$$\alpha_A = \dim \ker A, \quad \beta_A = \dim Z_A^*. \tag{1.89}$$

β_A is called the defect number and $\operatorname{ind} A = \chi_A = \beta_A - \alpha_A$ is the index of operator A. If $\alpha_A = 0$, then A establishes a one-to-one correspondence between $D(A)$ and $R(A)$. Hence, there exists $A^{-1} : R(A) \to D(A)$, $A^{-1}A\varphi = \varphi$ $[\varphi \in D(A)]$, and $AA^{-1}\psi = \psi$ $[\psi \in R(A)]$. In addition, A is a continuously invertible operator if $R(A) = Y$ and there exists a bounded inverse: $\|A\varphi\| \geq m\|\varphi\|, \psi \in R(A)$, and $m =$ const.

Definition 4 *If operator-valued function* $A(z)$ *is holomorphic at point* z_0 *and* $A(z)$ *is invertible at this point, then* z_0 *is called the regular point of operator* $A(z)$*. The multiplicity of any regular point is zero.*

Operator A is called normally solvable if the equation $A\varphi = \psi$, $\psi \in Y$, is solvable if and only if $f(\psi) = 0$ for every $f \in Z_A^*$. The latter condition is equivalent to the fact that $R(A)$ is closed.

Recall also that operator $A : X \to Y$ is called the closed operator if the conditions $\varphi_n \in D(A)$, $\varphi_n \to \varphi$, and $A\varphi_n \to \psi$ yield $\varphi \in D(A)$ and $A\varphi = \psi$.

Definition 5 *A linear closed normally solvable operator* A *with finite numbers* α_A *and* β_A *such that the index* $\chi_A = \beta_A - \alpha_A = 0$ *is called the* Φ*-operator.*

Let z_0 be a pole of operator-valued function $A(z)$. Then, in a vicinity of point z_0,

$$A(z) = \sum_{m=-M}^{\infty} (z - z_0)^m A_m. \tag{1.90}$$

If in expansion (1.90), A_m, $m = -1, -2, \ldots, -M$ are finite-dimensional operators, then operator-valued function $A(z)$ is called finite-meromorphic at point z_0. If $A(z)$ is finite-meromorphic at any point of a domain G, then it is called finite-meromorphic in G.

Note that the multiplicity of pole z_0 can be specified, as well as that of a CN, using Definition 3 in terms of the dimensions of the corresponding characteristic pole subspaces.

Definition 6 *Operator-valued function $A(z)$ is called Fredholm at a point z_0, if in expansion (1.90), the operator $A_0 = A(z_0)$ is the Φ-operator. If $A(z)$ is the Φ-operator at any point of a domain G, then it is called the Φ-operator in G.*

z_0 is called the normal point of operator-valued function $A(z)$ if $A(z)$ is finite-meromorphic and Fredholm at point z_0 and all points of a perforated vicinity of z_0 $0 < |z - z_0| < p$ for a certain $p > 0$ are regular points of $A(z)$.

Recall that the order of z_0 considered as a zero of vector-valued function $A(z)\varphi(z)$ is called the multiplicity of its root function $\varphi(z)$. The multiplicity (and order) is determined through the dimension $\alpha_A = \alpha_A(z)$ of the kernel of $A(z)$ at the point $z = z_0$. Similar statements are formulated for poles. We will not present the general definitions for the multiplicities of poles and CNs [one may refer, e.g., to [35, 50]; general definitions correspond to the cases when kernels and defect subspaces of $A(z)$ have arbitrary dimensions (finite or infinite)]. For the sake of brevity, we will limit the analysis to simple poles and CNs (that have unit multiplicities). For example, if z_0 is a simple CN of operator-valued function $A(z)$ and φ_0 is the corresponding eigenvector, then $\alpha_A(z_0) = 1$ and, according to Definition 3, $A(z)\varphi(z)$ can be represented in the form

$$A(z)\varphi(z) = (z - z_0)\Phi(z), \tag{1.91}$$

where $\Phi(z_0) \neq 0$ [i.e., z_0 is not a CN of operator-valued function $\Phi(z)$]. We note a particular case when the unit-multiplicity root function $\varphi(z)$ of $A(z)$ does not depend on z and coincides thus with eigenvector $\varphi_0 = \varphi(z_0)$. We will call such root functions (that correspond to simple CNs) the simple elements of operator-valued function $A(z)$. In the sequel, unless otherwise specified, we will consider simple poles and CNs (mainly of integral operator-valued functions).

Let $A(z)$ be a finite-meromorphic and Fredholm operator-valued function in domain G [i.e., every point $z \in G$ is the normal point of $A(z)$]; assume also that $A(z)$ is continuous up to the boundary $\Gamma = \partial G$. Then, $A(z)$ is called the normal operator-valued function with respect to contour Γ if $A(z)$ is invertible for all $z \in \Gamma \cap G$ except for a finite number of points in G that are the normal points of $A(z)$.

Assume that $A(z)$ is a normal operator-valued function with respect to contour Γ and $z_1, z_2, \ldots, z_n$ and $p_1, p_2, \ldots, p_m$ are, respectively, its simple CNs and poles that lie in G, and introduce the quantity

$$M[A(z), \Gamma] = n - m. \tag{1.92}$$

$M[A(z), \Gamma]$ may be called the total multiplicity of CNs and poles of operator-valued function $A(z)$ with respect to contour Γ (or domain G). Now, we can formulate three theorems, which play an important role in subsequent applications.

The first statement is known as the index perturbation theorem.

Theorem 11 (**[50]**) *If A is a Fredholm operator and B is a compact operator, then their indices coincide:* $\operatorname{ind}(A + B) = \operatorname{ind} A$.

The second statement is often referred to as the general analytical Fredholm theorem (see, for example, [37] or [50]) or the theorem about the discreteness of the spectrum of a Fredholm operator-valued function.

Theorem 12 **([35])** *Let $A(z)$ be a normal operator-valued function with respect to contour $\Gamma = \partial G$. Then, in domain G, there exists not more than a finite number of points $T_G = \{z_j\}_{j=1}^{n}$, at which*

$$\alpha_A(z) = \dim \ker A(z) > 0, \quad z \in T_G, \tag{1.93}$$

and

$$\alpha_A(z) = 0, \quad z \in G \backslash T_G. \tag{1.94}$$

In other words, the resolvent set $\rho_A(G)$ of operator-valued function $A(z)$ is $G \backslash T_A$, where its spectrum T_A consists of isolated points in G.

The next statement is the operator generalization of the Rouchet's theorem [35].

Theorem 13 **([35])** *Let $A(z)$ be a normal operator-valued function with respect to contour $\Gamma = \partial G$ with only simple poles and CNs in G. Let $S(z)$ be a finite-meromorphic operator-valued function in domain G continuous up Γ and satisfying the condition*

$$\|A^{-1}(z)S(z)\| < 1, \quad z \in \Gamma. \tag{1.95}$$

Then, $B(z) = A(z) + S(z)$ is a normal operator-valued function with respect to Γ and

$$M[B(z), \Gamma] = M[A(z), \Gamma]. \tag{1.96}$$

The integral operator-valued function

$$K(z, a_0)\varphi \equiv \int_{\Lambda} K(t_0, t; z)\varphi(t)dt = 0, \quad t_0 \in \Lambda, \tag{1.97}$$

where Λ is an interval (or a finite set of intervals), is called holomorphic in domain G if kernel $K(t_0, t; z)$ is an analytic function of z in G at each point $(t_0, t) \in \Pi = \Lambda \times \Lambda$. In this case, series (1.87) has the form

$$K(z) = K_0 + \sum_{m=1}^{\infty} (z - z_0)^m A_m, \tag{1.98}$$

where K_m are integral operators with the kernels

$$K_0(t_0, t) = K(t_0, t; z_0), \quad K_m(t_0, t) = \frac{1}{n!} \frac{\partial K^n(t_0, t; z)}{\partial z^n}\bigg|_{z=z_0}, \quad m = 1, 2, \ldots. \tag{1.99}$$

Let us proceed to the integral operator-valued function with a logarithmic singularity of the kernel

$$K(z)\varphi = [cL(a) + N(z)]\varphi \equiv \int_{-a}^{a} [c \ln |t_0 - t| + N(t_0, t; z)]\varphi(t)dt; \tag{1.100}$$

here, $c =$const, $N(t_0, t; z)$ is a smooth function satisfying conditions (1.14) and (1.16) for every z varying in a domain G of the complex plane, and $N(t_0, t; z)$ is a holomorphic functions of z in G for every point $(t_0, t) \in \overline{\Pi}_a = [-a, a] \times [a, a]$.

Following the results of Sections 1.1 and 1.2, one may consider K on the pair of Banach spaces $X = \tilde{H}_\mu$ and $Y = \tilde{H}^1_{\mu,0}$. Then, according to Theorem 3, $K : X \to Y$ is a Fredholm operator for every fixed $z \in G$ (because integral operator $N : X \to Y$ is completely continuous) and is a Fredholm holomorphic operator-valued function in G.

We will also consider a particular family of integral operator-valued functions (1.100) with a logarithmic singularity of the kernel, for which

$$N(t_0, t; z) = \frac{q_0(z) n_0(t_0) m_0(t)}{z - z_0} + N_0(t_0, t; z) \equiv \frac{A_{-1}(z)}{z - z_0} + N_0(t_0, t; z), \tag{1.101}$$

where N, N_0 (and q_0), and A_{-1} are holomorphic functions of z in the vicinity of z_0 for every point $(t_0, t) \in \overline{\Pi}_a$ and $n_0(t_0)$ and $m_0(t)$ are once continuously differentiable in $[-a, a]$. Here, again, it is easy to verify that $K : X \to Y$ given by (1.100) is a Fredholm operator for every fixed $z \neq z_0$ (because integral operators $N : X \to Y$ and $N_0 : X \to Y$ are completely continuous). Integral operator A_{-1} with a degenerated kernel is a one-dimensional operator. Therefore, according to definition (1.90), $K(z)$ is a Fredholm finite-meromorphic operator-valued function at point z_0. In addition, recalling the above statements, it is easy to see that z_0 is a simple pole of operator-valued function $K(z)$.

We see that integral operator-valued function $K(z)$ given by (1.100) and (1.101) can be represented as

$$K(z) = P_0(z) + N_0(z) \equiv cL(a) + \frac{A_{-1}}{z - z_0} + N_0(z). \tag{1.102}$$

where A_{-1} is a one-dimensional operator. Thus, it is natural to formulate the following

Definition 7 *We shall say that $K(z)$ given by (1.100) and (1.101) admits separation of the simple pole pencil $P_0(z)$ at a point z_0. If the kernels of $P_0(z)$ and $N_0(z)$ are real-valued functions for real values of z, we shall say that $K(z)$ admits separation of the simple-real-pole pencil.*

Theorem 14 *Let integral operator-valued function $K(z)$ admit separation of simple pole pencils at different points $Q_K = \{z_j,\ j = 1, 2, \ldots, M\}$ in a domain G of the complex plane z. Then, $K(z)$ is a Fredholm finite-meromorphic operator-valued function in G, and the resolvent set $\rho_K(G)$ is $G \backslash T_K$ where T_K is a set of isolated points in G.*

Theorem 14 may be considered as a corollary of Theorem 12 (or the general analytical Fredholm theorem [50]) as applied to this specific family of integral operator-valued functions with a logarithmic singularity of the kernel.

In order to prove Theorem 14, it is sufficient to verify that the resolvent set $\rho_K(G)$ of operator-valued function $K(z)$ under study is not empty. Indeed, using the explicit form of the operator $L^{-1}(a)$, one can prove that, for sufficiently small a and a fixed $z \neq z_i$,

$$\|L(a)^{-1} N(z)\|_X \leq C_K(a) < 1. \tag{1.103}$$

The complete proof of this statement is available in [72].

Estimate (1.103) can be extended to the integral operator-valued function $K(z)$ given by (1.100) and (1.101), which is holomorphic in domain $G \backslash Q_K$ of the complex plane z. Indeed, one can verify that $C_K(a) = C_K(a, z)$ in formula (1.103) admits, as a function of complex variable z, a uniform estimate $C_K(a, z) \leq C_0(a)$ on every compactum $G_0 \subset G$ that does not contain poles z_j for every $a \in (0, a_0)$ and a certain $a_0 > 0$. Hence, we may state that there exists a sufficiently small value of a such that every compactum $G_0 \subset G$ belongs to the resolvent set of $K(z)$.

In other words, for every fixed $z = z^* \neq z_i$, we may choose such a small a that there exists only a trivial solution of the equation

$$K(z^*)\varphi = 0,$$

[which, according to Definition 2, is the problem on CNs for operator-valued function $K(z)$] and, therefore, $z^* \in \rho_K(G)$ [since $\rho_K(G)$ is an open set]. We conclude that $\alpha(z) = \dim\ker K(z) = 0$, $z \in G\backslash T_K$, where $\alpha(z)$ denotes the dimension of the kernel of $K(z)$ at point z and T_K is a set of isolated points in G.

A more detailed analysis of integral operator-valued functions given by (1.100) and (1.101) can be performed using the properties of abstract pole pencils considered in Section 1.7 and the results of Section 1.8.

We complete this section by considering a finite-meromorphic integral operator-valued function $P_0(\lambda)$ with a logarithmic singularity of the kernel, which represents an example of a simple-real-pole pencil:

$$P_0(\lambda)\varphi = P_0(\lambda, a)\varphi = [L(a) + A_0(\lambda)]\varphi \equiv \int_{-a}^{a} \left[\frac{1}{\pi}\ln\frac{1}{|t_0 - t|} + \frac{b}{\lambda - \lambda_0}\right]\varphi(t)dt, \ t_0 \in (-a, a), \tag{1.104}$$

where λ is the spectral parameter and b and λ_0 are given numbers. Following the analysis of Section 1.2, one may consider $P_0 : X \to Y$ on the pair of Banach spaces $X = \tilde{H}_\mu$ and $Y = \tilde{H}^1_{\mu,0}$.

Let us demonstrate that in this particular case, we can determine CNs and eigenvectors of $P_0(\lambda)$ explicitly. Setting

$$\varphi(t) = \varphi_0(t) = \frac{1}{\sqrt{a^2 - t^2}}$$

and taking into account that

$$L(a)\varphi_0 = -\int_{-a}^{a} \frac{\ln|t_0 - t|}{\pi\sqrt{a^2 - t^2}}dt = \ln\frac{2}{a} = \text{const}$$

and

$$A_0(\lambda)\varphi_0 = \frac{b}{\lambda - \lambda_0}\int_{-a}^{a}\frac{dt}{\sqrt{a^2 - t^2}} = \frac{\pi b}{\lambda - \lambda_0} = \text{const},$$

we find that

$$P_0(\lambda, a)\varphi \equiv 0, \quad t_0 \in (-a, a)$$

at

$$\varphi = \varphi_0, \quad \lambda = \lambda^* = \lambda_0 + \frac{\pi b}{\ln\dfrac{2}{a}}.$$

That is, λ^* is a CN of $P_0(\lambda, a)$ and $\varphi_0 \in X$ is the corresponding eigenvector. In addition, $\lambda^* \to \lambda_0$ as $a \to 0$ and

$$|\lambda_0 - \lambda^*| = \frac{C}{|\ln a|}, \quad C = \text{const}.$$

Note that according to the terminology introduced above, φ_0 is a simple element of operator-valued function $P_0(\lambda, a)$ and λ_0 and λ^* are its simple pole and CN.

1.7 Abstract pole pencils and existence of characteristic numbers

Let us obtain the explicit formulas for the CN and for the inverse operator of the abstract pole pencil

$$P(\gamma) = \Upsilon + A_{-1}(\gamma-\gamma_0)^{-1} = \Upsilon + \langle ., \chi\rangle\psi(\gamma-\gamma_0)^{-1}, \quad (1.105)$$

where $A_{-1} : X \to R^1$ is a one-dimensional operator; Υ is an invertible operator on the pair of Banach spaces X, Y; $\langle ., \chi\rangle$ is a continuous functional on X; ψ is a given element of Y such that $u = \Upsilon^{-1}\psi \in X$ and $u \neq 0$; γ is a complex variable; and γ_0 is a given complex number. Note that in the case of integral operators (1.100) and (1.101) with a logarithmic singularity of the kernel, $\gamma = z$, Υ is operator $cL(a)$,

$$\psi = q_0 n_0(t_0), \quad \langle ., \chi\rangle = (., m_0) = \int_{-a}^{a} \cdot \cdot\, m_0(t)dt,$$

and the appropriate Banach spaces X and Y are specified in Sections 1.1 and 1.6. Below, when considering abstract operator-valued functions and operator pencils, we will use the notation γ for a complex variable.

We will consider the problem on CNs for the pole pencil

$$P(\gamma)\varphi = \Upsilon\varphi + \langle\varphi, \chi\rangle\psi(\gamma-\gamma_0)^{-1} = 0; \quad \varphi \in X, \;\; \psi, 0 \in Y. \quad (1.106)$$

Lemma 3 *Pole pencil $P(\gamma)$ has the unique CN γ^*. If $\langle u, \chi\rangle \neq 0$, then $\gamma^* \neq \gamma_0$.*

□ Taking into account the definition of u, we rewrite operator equation (1.106) in the equivalent form

$$\varphi + \langle\varphi, \chi\rangle u(\gamma-\gamma_0)^{-1} = 0. \quad (1.107)$$

One can substitute φ from (1.107) to (1.106) to obtain

$$\langle\varphi, \chi\rangle\psi + \langle(\gamma-\gamma_0)^{-1}\langle\varphi, \chi\rangle u, \chi\rangle\psi = 0. \quad (1.108)$$

Assume that $\gamma \neq \gamma_0$. Then, (1.108) is equivalent to

$$\langle\varphi, \chi\rangle[1 + \langle u, \chi\rangle(\gamma-\gamma_0)^{-1}] = 0. \quad (1.109)$$

If $\langle\varphi, \chi\rangle = 0$, then (1.106) has only the trivial solution, and none of the $\gamma \neq \gamma_0$ can be a CN.

Let us assume now that $\langle u, \chi\rangle \neq 0$; then

$$\gamma^* = \gamma_0 - \langle u, \chi\rangle \quad (1.110)$$

is the CN of $P(\gamma)$ with the corresponding eigenvector u □.

For the integral pole pencil

$$P_0(z) = cL(a) + \frac{A_{-1}}{z - z_0} \quad (1.111)$$

given by relationship (1.102), formula (1.110) for its CN $z^* = z_P^*$ takes the form

$$z_P^* = z_0 - \frac{q_0}{c}\langle L(a)^{-1}n_0, m_0\rangle. \quad (1.112)$$

Here, we have taken into account that q_0 is a nonzero constant.

Lemma 4 *The multiplicities of the pole and CN of the pole pencil $P(\gamma)$ each are equal to unity.*

□ We will prove this lemma only for the case of the pole multiplicity. Let us give an explicit form of the inverse operator $P^{-1}(\gamma)$. It is obvious that γ_0 is the pole of $P(\gamma)$ and the CN of $P^{-1}(\gamma)$. To represent the latter, consider the equation

$$\Upsilon\varphi + \frac{\langle\varphi,\chi\rangle\psi}{\gamma-\gamma_0} = f,$$

which is equivalent to

$$\varphi + \frac{\langle\varphi,\chi\rangle u}{\gamma-\gamma_0} = \Upsilon^{-1}f.$$

Substituting φ from this equation into the initial one and assuming that $\gamma \neq \gamma_0$, we obtain

$$\frac{\langle\varphi,\chi\rangle u}{\gamma-\gamma_0} = \frac{\langle\Upsilon^{-1}f,\chi\rangle u}{\gamma-\gamma_0} - \frac{\langle\varphi,\chi\rangle\langle u,\chi\rangle u}{(\gamma-\gamma_0)^2},$$

$$\langle\varphi,\chi\rangle = \frac{\langle\Upsilon^{-1}f,\chi\rangle}{1+(\gamma-\gamma_0)^{-1}\langle u,\chi\rangle} = \frac{\langle\Upsilon^{-1}f,\chi\rangle}{1-(\gamma-\gamma_0)^{-1}(\gamma^*-\gamma_0)} = \frac{(\gamma-\gamma_0)\langle\Upsilon^{-1}f,\chi\rangle}{\gamma-\gamma^*}.$$

One can see that $\gamma^* = \gamma_0$ only if $\langle u,\chi\rangle = 0$. Then,

$$\varphi = \Upsilon^{-1}f - (\gamma-\gamma^*)^{-1}\langle\Upsilon^{-1}f,\chi\rangle u,$$

and we obtain the required form

$$P^{-1}(\gamma)f = \Upsilon^{-1}f - (\gamma-\gamma^*)^{-1}\langle\Upsilon^{-1}f,\chi\rangle u, \tag{1.113}$$

where, in accordance with the above notations,

$$u = \Upsilon^{-1}\psi, \quad \gamma^* = \gamma_0 - \langle u,\chi\rangle,$$

and γ^* is the CN of $P(\gamma)$.

For integral pole pencil (1.111), formula (1.113) reduces to

$$P_0^{-1}(z)f = \frac{1}{c}L(a)^{-1}f - \frac{q_0}{c^2}\frac{1}{z-z_P^*}(L(a)^{-1}f, m_0)L(a)^{-1}n_0. \tag{1.114}$$

The statement of this lemma is now easily deduced from the explicit representations of operators P and P^{-1}.

Now, we can state sufficient conditions of the existence of CNs for finite-meromorphic integral operator-valued function $K(z)$ with a logarithmic singularity of the kernel defined by relationship (1.100).

Theorem 15 *Let finite-meromorphic integral operator-valued function $K(z)$ defined by (1.100) and (1.101) admit separation of simple pole pencils at different points $Q_K = \{z_j, j = 1,2,\ldots,M\}$ in a domain G of the complex plane z. Then, there exists a positive a_0 such that if $a \in (0,a_0)$, then there lies in the neighborhood of every point $z \in Q_K$ at least one CN of $K(z)$.*

□ Let $\mathcal{B}(z_0;p)$ be the circle $\{|z-z_0|=p\}$, where $z_0 \in Q_K$ with the radius less than the distance from z_0 to the closest point from Q_K [that is, to the neighboring pole of $K(z)$], i.e., $0 < p < \min_{z_0 \neq z \in Q_K} |z - z_0|$. We will use the same notation as that introduced above for the case of an abstract pole pencil, setting $\psi = q_0 n_0(t_0)$ ($q_0 = \text{const}$) and $\langle ., \chi\rangle = (., m_0)$, and take into account definition (1.100) and (1.101) of the considered integral operator-valued function $K(z)$. According to estimate (1.103),

$$\|L(a)^{-1}N_0(\gamma)\|_{X_a} \to 0, \quad \|L(a)^{-1}\psi\|_{X_a} \to 0,\ a \to 0$$

uniformly on every compactum $G_0 \subset G\backslash Q_K$. Applying the latter estimate to explicit formula (1.112) for the CN z_P^* of pole pencil $P_0(z)$ defined by (1.111) and taking into account the continuity of functions n_0 and m_0, we can easily verify that $z_P^* = z_P^*(a)$ tends to z_0 as $a \to 0$. Hence, for a sufficiently small $a_0 > 0$, for all positive values $a < a_0$, there are one pole and one CN of $P(z)$ in the circle $\mathcal{B}(z_0;p)$ with the corresponding radius $p = p(a_0)$, where p may be taken the same for all $a < a_0$. Therefore, the total multiplicity of $P(z)$ in $\mathcal{B}(z_0;p)$ is equal to zero. One can easily deduce from Lemma 3 with the help of formula (1.112) that for a fixed $a > 0$,

$$|z - z_P^*(a)| \geq \delta > 0, \quad z \in \mathcal{B}(z_0;p).$$

Applying explicit form (1.114) of the inverse $P_0^{-1}(z)$, we obtain the estimate

$$\sup_{z\in\mathcal{B}(z_0;p)} \|P^{-1}N_0(z)\|_{X_a} \leq$$

$$\sup_{z\in\mathcal{B}(z_0;p)} \|L(a)^{-1}N_0(z)\|_{X_a} + \sup_{z\in\mathcal{B}(z_0;p)} \frac{|(L(a)^{-1}N_0(z),\chi)| \|L(a)^{-1}\varphi\|_{X_a}}{|z-z_0|} = C_1(a) + C_2(a)p^{-1}.$$

According to estimate (1.103), the inequality

$$C_1(a) + C_2(a)p^{-1} < 1,$$

holds if a is sufficiently small. Applying Theorem 13 (the operator generalization of Rouchet's theorem [35]), one can easily verify that the total multiplicities of $K(z)$ and $P_0(z)$, corresponding to $\mathcal{B}(z_0;p)$, coincide. Hence, we conclude that they both are equal to zero, because the total multiplicity of $P(z)$ in $\mathcal{B}(z_0;p)$ is zero. Due to the above assumptions concerning the choice of the p-vicinity of z_0, $K(z)$ has one simple pole in $\mathcal{B}(z_0;p)$, and we finally obtain that, in $\mathcal{B}(z_0;p)$, operator-valued function $K(z)$ has at least one CN. □

1.8 Logarithmic integral operators in Sobolev spaces

In this section, we summarize the main results concerning the properties of logarithmic operators in the Sobolev spaces.

First, we focus on a mathematical statements employing the properties of the single-layer potentials on curves Γ and consider logarithmic equations on these curves. The logarithmic equation is reduced to a pseudodifferential operator (PDO) equation on $\overline{\Gamma}$. The theory is constructed in the Sobolev spaces. The equations on rectilinear intervals constitute an important particular case when the spaces of solutions and images and PDOs can be described in terms of the Fourier transform.

In the general case of an arbitrary curve, the logarithmic equation is also reduced to the analysis of a PDO acting on manifold $\overline{\Gamma}$. We separate the principal part of the corresponding operator in the local coordinates and use the matching procedure to construct the PDO that differs from the initial operator by an operator with an infinitely smooth kernel. We prove that the PDO considered as an operator acting on the pair of Sobolev spaces is a bounded and Fredholm operator with zero index. The logarithmic equation is considered as a pseudodifferential equation, and the generalized solution is defined.

Consider a system composed of a finite number of simple, planar, closed or unclosed smooth curves Γ_j of a finite length that belongs to class C^∞. We introduce the following notation:

$$\Gamma = \bigcup_j \Gamma_j, \quad \overline{\Gamma}_i \bigcap \overline{\Gamma}_j = \emptyset \text{ for } \quad i \neq j,$$

where

$$\partial\Gamma = \bigcup_j (\overline{\Gamma}_j \setminus \Gamma_j)$$

are the endpoints of Γ (these points do not belong to Γ, $\partial\Gamma \bigcap \Gamma = \emptyset$). Let G_0 be a union of internal domains bounded by closed curves Γ_j, and we set $G_0 = \emptyset$ if these domains are absent.

Let Λ be a closed, smooth C^∞-curve that contains Γ ($\Gamma \subset \Lambda$) and divides the plane into bounded, G_1, and unbounded, G_2, domains. We will assume that n is the normal vector external with respect to G_1.

Let us formulate the definitions of the Sobolev spaces. Consider Λ as a closed manifold and denote by $U = \{U_\alpha\}$ a finite cover of Λ by the coordinate vicinities and by $\{\varphi_\alpha\}$, the partition of unity subordinated to the cover of U. For every $v \in C^\infty(\Gamma)$, we introduce the functions $v_\alpha = \varphi_\alpha v$ defined in local coordinates in a certain interval of the real line and set

$$\|v\|_s^2 = \sum_\alpha \|v_\alpha\|_s^2, \quad s \in \mathrm{R}$$

for every $v \in C^\infty(\Lambda)$. The norm in the space $H^s(\mathrm{R}^1)$ is defined as

$$\|v_\alpha\|_s^2 = \int_{-\infty}^{\infty} \left(1 + |\xi|^2\right)^s |\widehat{v}_\alpha(\xi)|^2 \, d\xi,$$

where $\widehat{v}_\alpha$ is the Fourier transform of function v_α. By $H^s(\Lambda)$ we will denote the completion of $C^\infty(\Lambda)$ with respect to the $\|\cdot\|_s$-norm. Note that if the coordinate vicinities and the partition of unity are chosen in a different manner, one obtains an equivalent norm; therefore, the definition of $H^s(\Lambda)$ is correct.

Let $\Gamma \subset \Lambda$ be a submanifold of Λ (a part of curve Λ); we set

$$H^s(\Gamma) = \{v|_\Gamma : v \in H^s(\Lambda)\}$$

and

$$\tilde{H}^s(\Gamma) = \left\{v \in H^s(\Lambda) : \operatorname{supp} v \subset \overline{\Gamma}\right\}.$$

Spaces $H^s(\Gamma)$ and $\tilde{H}^{-s}(\Gamma)$ are antidual, for all $s \in \mathrm{R}$, with respect to sesquilinear form $\int_\Gamma v\overline{w}\, dl$. The space $\tilde{H}^s(\Gamma)$ may be constructed as a closure of $C_0^\infty(\Gamma)$ with respect to the norm $\|\cdot\|_s$. For a detailed description of the properties of spaces $H^s(\Lambda)$, $H^s(\Gamma)$, and $\tilde{H}^s(\Gamma)$, one can refer, for example, to [64, 65].

Let us introduce operators γ_0 of the trace on Λ; γ_1, of the trace of the normal derivative on Λ (on each side of the curve); q, of the zero continuation of a function from Γ to Λ; and p, of the restriction of a function from Λ to Γ:

$$\gamma_0 : \; u \to u|_\Lambda : H^1_{loc}(\mathrm{R}^2) \to H^{1/2}(\Lambda),$$

$$\gamma_1 : u \to \left.\frac{\partial u}{\partial n}\right|_\Lambda : H^1_P(G) \to H^{-1/2}(\Lambda),$$

$$q : \varphi \to \tilde{\varphi} : \tilde{H}^s(\Gamma) \to H^s(\Lambda),$$

$$p : \varphi \to \varphi|_L : H^s(\Lambda) \to H^s(\Gamma),$$

where $G = G_1$ or $G = G_2$ and $H^1_P(G) = \{u \in H^1(G) : \Delta u \in L_2(G)\}$. According to [11], all operators are continuous on the corresponding pair of spaces; therefore, it is sufficient to define them for smooth functions (in an obvious manner) and to extend then to the whole space using continuation.

We will consider the single-layer potential

$$u(\mathbf{x}) = S_0(q\varphi) = -\frac{1}{2\pi}\int_\Gamma \ln\frac{1}{|\mathbf{x}-\mathbf{y}|}\varphi(\mathbf{y})\,dl, \tag{1.115}$$

$$\varphi \in \tilde{H}^{-1/2}(\Gamma), \quad \mathbf{x}, \mathbf{y} \in \mathrm{R}^2.$$

For the case of a closed contour, potential (1.115) is investigated in [11]. Continuing φ as zero functions from Γ to Λ and then again performing the restriction on Γ, we obtain, by virtue of the continuity of operators q and p and the results of [11], that

$$\varphi = \left[\frac{\partial u}{\partial n}\right]_\Gamma \quad \text{and} \quad [u]_\Gamma = 0.$$

According to [11],

$$u \in H^1_{loc}(\mathrm{R}^2 \setminus \overline{G_0})$$

for all

$$\varphi \in \tilde{H}^{-1/2}(\Gamma)$$

in (1.115). Here,

$$[u]_\Gamma = p\left(\gamma_0\left(u|_{G_2}\right) - \gamma_0\left(u|_{G_1}\right)\right),$$

$$\left[\frac{\partial u}{\partial n}\right]_\Gamma = p\left(\gamma_1\left(u|_{G_2}\right) - \gamma_1\left(u|_{G_1}\right)\right).$$

Solving the Dirichlet boundary value problem for the Laplace equation in $R^2 \setminus (\overline{\Gamma} \bigcup G_0)$, we can find the solution in the form of single-layer potential (1.115). This yields the equation on curves Γ:

$$L\varphi = \frac{1}{2\pi}\int_\Gamma \ln\frac{1}{|\mathbf{x}-\mathbf{y}|}\varphi(\mathbf{y})\,dl = f(\mathbf{x}), \quad \mathbf{x} \in \Gamma, \tag{1.116}$$

$$L = p\gamma_0 S_0 q.$$

The operator of this equation has a weak (logarithmic) singularity. (1.116) may be considered as a pseudodifferential equation.

Before to proceed to the analysis of the general case of the problem on Γ, we consider a particular case of a rectilinear interval, when

$$\Gamma = \{\mathbf{x} = (x, y) : y = 0, a < x < b, \}.$$

In this case, L is a convolution-type operator on line R^1. In Section it was shown that L is not injective on segment $[-2, 2]$. Hence, we cannot prove the bijectivity of L for arbitrary Γ. We will prove only the Fredholm property of operator L.

With this purpose, we will look for solutions φ in the spaces of distributions $\varphi \in \tilde{H}^{-1/2}(\Gamma)$ and consider the operator

$$L : \tilde{H}^{-1/2}(\Gamma) \to H^{1/2}(\Gamma).$$

In the case under consideration, these spaces can be described in terms of the Fourier transform:

$$H^s(\mathrm{R}^1) = \left\{v : (1 + |\xi|)^s \hat{v}(\xi) \in L_2(\mathrm{R}^1), \right\},$$

$$\tilde{H}^s(\Gamma) = \left\{v \in H^s(\mathrm{R}^1), \operatorname{supp} v \subset \overline{\Gamma}\right\},$$

and $H^s(\Gamma)$ is a restriction of $H^s(\mathrm{R}^1)$ on Γ. Here, $\hat{v}$ denotes the Fourier transform of v.

Denote

$$K(x) = \ln \frac{1}{\sqrt{x^2 + 1}}.$$

Since $K(|x_0 - x|) \in C^\infty(\mathrm{R}^2)$ [recall that $\mathbf{y} = (x_0, y_0)$], the operator

$$K\varphi = \frac{1}{2\pi} \int\limits_\Gamma K(|x_0 - x|)\varphi(x)\, dx, \tag{1.117}$$

$$K : \tilde{H}^{-1/2}(\Gamma) \to H^{1/2}(\Gamma)$$

is compact.

Consider the operator $\tilde{L}$

$$\tilde{L}\varphi = \frac{1}{2\pi} \int\limits_\Gamma \left(\ln \frac{1}{|x_0 - x|} - K(|x_0 - x|) \right) \varphi(x)\, dx.$$

L and $\tilde{L}$ are Fredholm operators with the zero index.

From [55], it follows that

$$\frac{1}{\pi} \int\limits_{-\infty}^{+\infty} e^{-ix\xi} \ln \frac{\sqrt{x^2 + 1}}{|x|}\, dx = \frac{1}{|\xi|}(1 - e^{-|\xi|}).$$

Using the convolution theorem and the properties of the Fourier transform, we can represent $\tilde{L}$ as a PDO

$$\tilde{L}\varphi = \int\limits_{-\infty}^{+\infty} \frac{1}{|\xi|}(1 - e^{-|\xi|}) e^{ix\xi} \hat{\varphi}(\xi)\, d\xi, \quad x \in \Gamma. \tag{1.118}$$

We have

$$\frac{1}{|\xi|}(1 - e^{-|\xi|}) \sim |\xi|^{-1} \quad \text{as} \quad |\xi| \to \infty;$$

therefore, according to [66], $\tilde{L}$ is a Fredholm PDO with the zero index and order -1. This operator is injective by virtue of the estimate

$$\left(\tilde{L}\varphi, \varphi\right)_{L_2} \geq \int\limits_{-\infty}^{+\infty} |\xi|^{-1}(1 - e^{-|\xi|})|\hat{\varphi}(\xi)|^2 \, d\xi > 0, \quad \varphi \neq 0.$$

This estimate proves that the kernel of the operator contains only the zero element: $\ker \tilde{L} = \{0\}$. Thus, according to the Fredholm alternative, operator (1.118) is bounded and has a bounded inverse. Hence, L is a Fredholm operator with the zero index, because $L = \tilde{L} + K$, where $\tilde{L}$ has a bounded inverse and K is a compact operator.

Consider the general case of a set of curves Γ of arbitrary shape. We will consider a manifold with edge $\overline{\Gamma}$ as a submanifold Λ. Note that it is not necessary to choose a specific cover of $\overline{\Gamma}$ and a specific coordinate system in the vicinity of edge $\partial\Gamma$.

Let Λ be a closed, connected, oriented C^∞ curve (note that a closed curve is a one-dimensional compact manifold without edge). Let $\overline{\Gamma} \subset \Lambda$ be a submanifold with edge of a manifold Λ, which is not necessarily connected and has a finite number of connected components; each of these components has the dimension one.

We fix a finite cover $U = \{U_\alpha\}$ of Λ by coordinate vicinities and denote by $\chi_\alpha : U_\alpha \to V_\alpha \subset \mathrm{R}^2$ the local maps and by $\{\varphi_\alpha\}$, a partition of unity subordinated to U. For a function $v \in C^\infty(\Lambda)$, we have $\varphi_\alpha v \in C_0^\infty(V_\alpha)$.

Now we will consider operator $\tilde{L}$ on curve Γ:

$$\tilde{L}\varphi = \frac{1}{2\pi} \int\limits_\Gamma \left(\ln \frac{1}{|\mathbf{x} - \mathbf{y}|} - K(|\mathbf{x} - \mathbf{y}|) \right) \varphi(\mathbf{y}) \, dl.$$

L and $\tilde{L}$ are Fredholm operators with zero index.

Consider the kernel of integral operator $\tilde{L}$:

$$\ln\left(1 + \frac{1}{|\mathbf{x} - \mathbf{y}|^2}\right).$$

Fix α, $U = U_\alpha$, and $V = V_\alpha$. Let $\chi^{-1} : V \to U$ and $\mathbf{x} = \chi^{-1}(x) \in U$, where $x \in V$ are local coordinates on U. We have

$$|\mathbf{x} - \mathbf{y}| = |\chi^{-1}(x) - \chi^{-1}(x_0)| = \Phi(x, x_0)|x - x_0|,$$

where $\Phi(x, x_0) \in C^\infty(V \times V)$ (since U is C^∞ curve and $\Phi(x, x_0) > 0$ for $x, x_0 \in V$; in addition to this, $\Phi(x, x_0) = \Phi(x_0, x)$. We set

$$\Phi(x, x_0) = \Theta(x_0) + \Pi(x, x_0), \quad \Theta(x_0) = \Phi(x_0, x_0), \quad \Pi \in C^\infty(V \times V),$$

where $\Pi(x_0, x_0) = 0$ for $x_0 \in V$. Then

$$\begin{aligned}
&\ln\left(1 + \frac{1}{|\mathbf{x} - \mathbf{y}|^2}\right) = \ln\left(1 + \frac{1}{\Phi^2(x, x_0)|x - x_0|^2}\right) \\
= \; &\ln \frac{|x - x_0|^2 + \Theta^{-2}(x_0)}{|x - x_0|^2} + \ln \frac{1}{\Theta^2(x_0)} + \ln \frac{\Theta^2(x_0)\left(\Phi^2(x, x_0)|x - x_0|^2 + 1\right)}{\Phi^2(x, x_0)\left(\Theta^2(x_0)|x - x_0|^2 + 1\right)} \\
= \; &\ln \frac{|x - x_0|^2 + \Theta^{-2}(x_0)}{|x - x_0|^2} + \Psi(x, x_0),
\end{aligned}$$

where $\Psi \in C^\infty(V \times V)$.

On curve U, we use the natural parameter as a local coordinate, so that $dl = dx$.

Depending on situation, we will consider operator $\tilde{L}$ (which acts in a functional space) as a PDO on manifolds Λ or Γ. For every coordinate vicinity $U = U_\alpha$, and, respectively, $V = V_\alpha$, we define the restriction $\tilde{L}$ on V by the formula

$$\tilde{L}_V = p_V \tilde{L} q_V : C_0^\infty(V) \to C^\infty(V),$$

where $q_V : C_0^\infty(V) \to C^\infty(\Lambda)$ is a natural inclusion (the zero continuation outside V), $p_V : C^\infty(\Lambda) \to C^\infty(V)$ is the operator of restriction that transforms f to $f|_V$. If $\tilde{L}_V$ is transformed into a PDO, then $\tilde{L}$ is a PDO on manifold Λ [70, 69] or Γ if to consider the restriction of $\tilde{L}$ on Γ:

$$\tilde{L} : C_0^\infty(\Gamma) \to C^\infty(\Gamma).$$

Let us define operator $\tilde{L}$ on $C_0^\infty(\Gamma)$. Since it is necessary to know only the values $\tilde{L}\varphi$ at points $\mathbf{x} \in \Gamma$ (or $\mathbf{x} \in \Lambda$), we will use the following representation for $\tilde{L}_V$:

$$\tilde{L}_V \varphi = \int_V \ln\left(1 + \frac{\Theta^{-2}(x_0)}{|x - x_0|^2}\right) \varphi(x)\, dx \tag{1.119}$$

$$+ \int_V \Psi(x, x_0) \varphi(x)\, dx. \tag{1.120}$$

The first term in (1.120) is a convolution-type integral operator. Let us calculate the Fourier transform of the kernel of this operator (see [55]):

$$\frac{1}{\pi} \int_{-\infty}^{+\infty} e^{-ix\xi} \ln \frac{\sqrt{x^2 + \Theta^{-2}(x_0)}}{|x|}\, dx = \frac{1}{|\xi|}\left(1 - e^{-\frac{|\xi|}{\Theta(x_0)}}\right). \tag{1.121}$$

Then, (1.120) can be rewritten as

$$\tilde{L}_V \varphi \equiv L_V^0 \varphi + K_V \varphi = \int_{-\infty}^{+\infty} \frac{1}{|\xi|}\left(1 - e^{-\frac{|\xi|}{\Theta(x_0)}}\right) e^{ix\xi} \widehat{\varphi(\xi)}\, d\xi + \int_V \Psi(x, x_0) \varphi(x)\, dx, \tag{1.122}$$

where $\widehat{\varphi}(\xi)$ is the Fourier transform of function $\varphi \in C_0^\infty(V)$ and $\Psi \in C^\infty(V \times V)$. Formula (1.122) defines the PDO

$$L_V^0 : C_0^\infty(V) \to C^\infty(V)$$

of the order -1 with the positively homogeneous (with respert to ξ) principal symbol $|\xi|^{-1}$.

The last term in (1.120) and (1.122) produces the operator with an infinitely smooth kernel, so that operator K_V belongs to $L^{-\infty}(V)$ [69, 70].

Consider operator $\hat{L}$ defined by the formula

$$\hat{L}\varphi = \sum_\alpha \psi_\alpha L_\alpha^0 \varphi_\alpha \varphi, \tag{1.123}$$

where $L_\alpha^0 \equiv L_V^0$ for $V = V_\alpha$, $1 = \sum_\alpha \varphi_\alpha$, and functions ψ_α are such that $supp\, \psi_\alpha \subset V_\alpha$ and $\psi_\alpha \varphi_\alpha \equiv \varphi_\alpha$. The operator

$$\hat{L} : C_0^\infty(\Gamma) \to C^\infty(\Gamma).$$

Since the kernel of $\tilde{L}$ has a singularity only at $x = y$, operator $\tilde{L}$ differs from $\widehat{L}$ by an operator with an infinitely smooth kernel:

$$\tilde{L} = \widehat{L} + \widehat{K}, \quad \widehat{K} \in L^{-\infty}(\Gamma). \tag{1.124}$$

The PDO with the symbol $|\xi|^{-1}\left(1 - e^{-|\xi|/\Theta(x_0)}\right)$ defines an isomorphic mapping of $H^{-1/2}(\Lambda)$ on $H^{1/2}(\Lambda)$ [69, 70]. $\Theta(x_0) > 0$ in every coordinate vicinity, $C_0^\infty(\Gamma)$ is dense in $\tilde{H}^{-1/2}(\overline{\Gamma})$, and $\tilde{H}^{-1/2}(\overline{\Gamma})$ is antidual with respect to $H^{1/2}(\Gamma)$; therefore, the definition of operator $\widehat{L}$ can be extended to a bounded, continuously invertible operator

$$\widehat{L}: \ \tilde{H}^{-1/2}(\overline{\Gamma}) \to H^{1/2}(\Gamma).$$

Then, by virtue of (1.124),

$$\tilde{L}: \ \tilde{H}^{-1/2}(\overline{\Gamma}) \to H^{1/2}(\Gamma)$$

is a bounded Fredholm operator with zero index.

Thus, we have proved the following statement.

Theorem 16

$$L: \ \tilde{H}^{-1/2}(\Gamma) \to H^{1/2}(\Gamma)$$

is a bounded Fredholm operator with zero index.

Note that the Gårding inequality holds

$$((L + K_L)\varphi, \varphi)_{L_2} \ \geq \ \lambda_L \|\varphi\|^2_{-1/2}, \quad \forall \varphi \in \tilde{H}^{-1/2}(\Gamma),$$

with a certain compact operator

$$K_L \ : \ \tilde{H}^{-1/2}(\Gamma) \to H^{1/2}(\Gamma)$$

and $\lambda_L > 0$.

Assume that all curves Γ_j are closed. Then,

$$L: \ H^{-1/2}(\Gamma) \to H^{1/2}(\Gamma)$$

is a Fredholm operator with zero index, where L is an elliptic PDO of the order -1. Equation (1.125) [or (1.116)] is uniquely solvable if Γ are nonresonance curves [62, 67].

Using the regularity theorem for elliptic PDOs [69, 70], we can easily verify that if the equation

$$L\varphi = f \tag{1.125}$$

has the $C^\infty(\overline{\Gamma})$ right-hand side and Γ is a C^∞ curve, then solution φ on Γ belongs to $C^\infty(\Gamma)$.

It is important to investigate the asymptotic behaviour of solutions in the vicinities of endpoints of $\partial\Gamma$. Let f be smooth functions, e.g., $f \in C^\infty(\overline{\Gamma})$. Applying the known results concerning the regularity of solutions of elliptic equations, it is easy to show [68] that φ are smooth functions on Γ (they belong to C^∞ if all $\Gamma_j \in C^\infty$) and have the singularities

$$\varphi = O(\rho^{-1/2}), \quad \rho \to 0, \tag{1.126}$$

in the vicinities of the endpoints of $\partial\Gamma$ (ρ is the distance to an endpoint of the curve); note that estimates (1.126) are accurate with respect to the order.

1.9 Integral operator-valued functions with the kernels represented by series with meromorphic coefficients

Consider the case when the kernel of an integral operator-valued function $K = K(\gamma) : X \to Y$ is represented in the form of a series with meromorphic coefficients:

$$K(\gamma)\varphi = \int_{-a}^{a} K(t_0, t; \gamma)\varphi(t)dt, \quad K(t_0, t; \gamma) = \sum_{n=0}^{\infty} \beta_n(\gamma)\psi_n(t_0)\psi_n(t), \tag{1.127}$$

where the system of functions $\{\psi_n\}$ is complete in Hilbert space H with inner product $(\cdot, \cdot)$, such that $X \subset Y$ and $Y \subset H$. The formal operator series

$$\sum_{n=0}^{\infty} \beta_n(\gamma)(., \psi_n)\psi_n$$

is assumed to converge in the operator norm, and coefficients $\beta_n(\gamma)$ are meromorphic functions in domain G, so that $K(\gamma)$ is a finite-meromorphic operator-valued function.

For this type of operator-valued functions, the following theorem can be used for establishing the fact that the resolvent set is not empty.

Theorem 17 ([40]) *Under the conditions*

$$\beta_n(\gamma_0) < 0 \quad [\beta_n(\gamma_0) > 0], \ n = 0, 1 \ldots$$

point γ_0 belongs to the resolvent set $\rho_K(G)$.

□ To prove this theorem, it is sufficient to apply the above conditions and to verify that the quadratic form

$$F_K(\gamma) = (K(\gamma)\varphi, \varphi) = \sum_{n=0}^{\infty} \beta_n(\gamma)|(\varphi, \psi_n)|^2$$

is of fixed sign. □

Theorem 17 does not guarantee the existence of CNs. This can be proved in some special situations. Below, we will consider two such cases.

Assume that

(i) $\{\psi_n\}_{n=0}^{\infty}$ form a complete system of trigonometric or exponential functions in $L^1(a, b)$, not necessarily orthogonal (for example, the system $\psi_n = \cos \pi n x, n = 0, 1, 2, \ldots$ is complete but not orthogonal in $L^1(a, b)$ if $[a, b] \subset [0, \pi]$);

(ii) for any fixed γ, $\beta_n(\gamma) = O(k^{-1})$, as $k \to \infty$;

(iii) there exists a real γ^* such that for all real $\gamma < \gamma^*$, all $\beta_n(\gamma)$ are real and $\beta_n(\gamma) > 0$;

(iv) each function $\beta_n(\gamma)$ has a countable set of real poles on the real axis with the accumulation point at $+\infty$;

(v) $\frac{d}{d\gamma}\beta_n(\gamma) = O(k^{-2})$, as $k \to \infty$;

(vi) at all regular points, $\frac{d}{d\gamma}\beta_n(\gamma) > 0, \ n = 0, 1, 2, \ldots$.

Theorem 18 ([40]) *Under conditions* (i)–(vi), *operator-valued function* $K(\gamma)$ *has a countable number of real CNs with the accumulation point at* $+\infty$. *Between two neighbouring poles of each function* β_n, *there is at least one CN.*

□ As follows from conditions (i)–(iii), for any $\gamma < \gamma^*$ and any φ,

$$(K(\gamma)\varphi, \varphi) = \sum_{n=0}^{\infty} \beta_n(\gamma)|(\varphi, \psi_n)|^2 \geq 0.$$

Moreover, the equality to zero is attained only for $\varphi \equiv 0$ due to the completeness of system $\{\psi_n\}$. Hence, the resolvent set $\rho(K)$ is not empty and, therefore, the inverse operator-valued function $K^{-1}(\gamma)$ exists.

Let γ_1 and γ_2 be two neighbouring real poles of the same function $\beta_n(\gamma)$. Since these points are poles of operator-valued function $K(\gamma)$, they are CNs of $K^{-1}(\gamma)$. Moreover, they correspond to the same root function ψ_n.

Let us consider the function $F(\gamma) = (K^{-1}(\gamma)\psi_n, \psi_n)$, which is obviously meromorphic. Since ψ_n is a root function of $K^{-1}(\gamma)$, function $F(\gamma)$ vanishes at the ends of interval $[\gamma_1, \gamma_2]$.

Denote $\varphi(\gamma) = K^{-1}(\gamma)\psi_n$. Then,

$$\frac{d}{d\gamma}(K^{-1}(\gamma)\psi_n, \psi_n) = \left(\frac{d}{d\gamma}(K^{-1}(\gamma)\psi_n), \psi_n\right)$$

$$= (\varphi'(\gamma), K(\gamma)\varphi(\gamma)) = (K(\gamma)\varphi'(\gamma), \varphi(\gamma)).$$

Differentiating the equality

$$K(\gamma)\varphi(\gamma) = \psi_n,$$

we obtain

$$K(\gamma)\varphi'(\gamma) = -K'(\gamma)\varphi(\gamma).$$

Then, taking into account conditions (v) and (vi), we have

$$\frac{d}{d\gamma}(K^{-1}(\gamma)\psi_n, \psi_n) = -(K'(\gamma)\varphi(\gamma), \varphi(\gamma)) = -\sum_{n=0}^{\infty} \beta_n'(\gamma)|(\varphi(\gamma), \psi_n)|^2 \geq 0, \quad \gamma \in (\gamma_1, \gamma_2).$$

The equality $(\varphi(\gamma), \psi_n) = 0$ for all n yields $\varphi(\gamma) = 0$ (as an element of the functional space) due to the completeness of system $\{\psi_n\}$. The corresponding γ are CNs of $K^{-1}(\gamma)$; therefore, they are poles of $K(\gamma)$, and according to condition (iv), only a finite number of them fall into (γ_1, γ_2).

Thus, $F'(\gamma) > 0$ for $\gamma \in (\gamma_1, \gamma_2)$ with possible exception of a finite number of points at which $F'(\gamma) = 0$. Therefore, function $F(\gamma)$ monotonically increases in the interval (γ_1, γ_2), and, since $F(\gamma)$ vanishes at the ends of this interval, this function must have break points in (γ_1, γ_2). $F(\gamma)$ is a meromorphic function; therefore, these points coincide with poles of $F(\gamma)$; consequently, they are poles of $K^{-1}(\gamma)$ and CNs of $K(\gamma)$. □

Theorem 18 can be extended to a wider class of abstract operator-valued functions repesented by series.

1.10 Methods of small parameter: approximate inversion and semi-inversion

In this section, we focus on some useful techniques [42, 45] that can be applied to problems formulated in terms of logarithmic integral equations. If the set of integration is small (in a certain sense), the integral operator can be represented (at least locally) as a sum of an invertible operator and a finite-dimensional operator. The former can be inverted using an expansion in the small parameter—the length of the set of integration or the inverse logarithm of this value. Then, the operator of the problem can be inverted, which enables one to obtain approximate solution to the logarithmic integral equation or to find CNs of logarithmic integral operators.

In this section, we impose stronger requirements on the completely continuous part $N(\gamma)$ of operator-valued function $K(\gamma)$ than those specified by (1.14) and (1.15). We assume that kernel $N(x_0, x; \gamma)$ admits the following estimate for $x_0 - x \to 0$:

$$|N(x_0, x, \gamma) - N(0, 0, \gamma)| \leq C_1(\gamma)(x_0 - x)^2 \ln \frac{1}{|x_0 - x|} + C_2(\gamma)x_0^2 + C_3(\gamma)x^2, \quad (1.128)$$

where $C_1(\gamma)$, $C_2(\gamma)$, and $C_3(\gamma)$ are constants independent of x and x_0. These conditions are fulfilled in a wide class of boundary value problems for example, related to the wave diffraction by strips and slots, free oscillations in closed and open resonators, guided waves in screened and open structures, etc. We also assume that $N(x_0, x, \gamma)$ is analytical as a function of γ in the whole complex plane with an exception of a set of isolated singular points (first-order poles and logarithmic branch points).

1.10.1 Approximate inversion

Consider the integral equation

$$K(\gamma)\varphi \equiv \int_{-l}^{l} \left[\alpha \frac{1}{\pi} \ln \frac{1}{|x_0 - x|} + N(x_0, x, \gamma)\right] \varphi(x) dx = f(x_0), \quad x_0 \in [-l, l], \quad \alpha = \text{const}.$$

Performing the change of variables

$$x = lt, \quad x_0 = lt_0, \quad (1.129)$$

we obtain the equivalent integral equation

$$\tilde{K}(\gamma, l)\tilde{\varphi} = \tilde{f}, \quad (1.130)$$

where $\tilde{K}(\gamma, l)$ is the transformed operator-valued function

$$\tilde{K}(\gamma, l)\tilde{\varphi} = \int_{-1}^{1} \left[\alpha \frac{1}{\pi} \ln \frac{1}{|t_0 - t|} + \alpha \frac{1}{\pi} \ln \frac{1}{l} + N(lt_0, lt, \gamma)\right] \tilde{\varphi}(t) dt, \quad t_0 \in [-1, 1],$$

$\tilde{\varphi}(t) = l\varphi(lt)$, and $\tilde{f}(t_0) = f(lt_0)$. The use of $\tilde{K}(\gamma, l)$ instead of $K(\gamma)$ is more convenient, because the set of integration and, therefore, the functional spaces in which $\tilde{K}(\gamma, l)$ acts are the same for all values of l. The dependence on l is transferred into the kernel of the integral operator, which facilitates the analysis. Evidently, the spectra of operator-valued functions $K(\gamma)$ and $\tilde{K}(\gamma,l)$ coincide.

As follows from estimate (1.128), $\tilde{K}(\gamma,l)$ can be represented in the form

$$\tilde{K}(\gamma,l)\tilde{\varphi}=\alpha L\tilde{\varphi}+\alpha\frac{1}{\pi}\ln\frac{1}{l}(\tilde{\varphi},1)1+M_0(\gamma)(\tilde{\varphi},1)1+l^2\ln lR(\gamma,l)\tilde{\varphi},$$

where $M_0(\gamma)=M(0,0,\gamma)$ and $R(\gamma,l)$ is an operator-valued function holomorphic with respect to γ and uniformly bounded as $l\to 0$. Let us introduce operator-valued function $L_1(\gamma,\beta)$ defined by the equality

$$L_1(\gamma,\beta)\tilde{\varphi}=\alpha L\tilde{\varphi}+\left(\frac{\alpha}{\beta}+M_0(\gamma)\right)(\tilde{\varphi},1)1,$$

where

$$\beta=\left(\frac{1}{\pi}\ln\frac{1}{l}\right)^{-1}.$$

Inverse operator-valued function $L_1^{-1}(\gamma,\beta)$ is given by the formula

$$L_1^{-1}(\gamma,l)f=\frac{1}{\alpha}L^{-1}f-\frac{1}{\alpha}\frac{\left(\frac{1}{\beta}+\frac{M_0(\gamma)}{\alpha}\right)(L^{-1}f,1)}{1+\frac{\pi}{\ln 2}\left(\frac{1}{\beta}+\frac{M_0(\gamma)}{\alpha}\right)}L^{-1}1$$

where operator L^{-1} is defined by (1.5).

If β is sufficiently small, then L^{-1} can be expanded in powers of β:

$$L_1^{-1}(\gamma,\beta)f=\frac{1}{\alpha}L^{-1}f$$

$$-\frac{\ln 2}{\alpha\pi}\left(1-\beta\frac{\ln 2}{\pi}+\beta^2\frac{\ln 2}{\pi}\left(\frac{M_0(\gamma)}{\alpha}+\frac{\ln 2}{\pi}\right)\right)(L^{-1}f,1)L^{-1}1+O(\beta^3),\quad\text{as}\quad\beta\to 0. \tag{1.131}$$

Lemma 5 *For any fixed γ, there exist a value β_0 and a constant $C>0$ such that, for any $\beta<\beta_0$, the estimate $\|L_1^{-1}(\gamma,\beta)\|<C$ is valid.*

□ As follows from representation (1.131),

$$\|L_1^{-1}f\|\le\left|\frac{1}{\alpha}\right|\|L^{-1}\|\|f\|+C_1\left|\frac{\left(\frac{1}{\beta}+\frac{M_0(\gamma)}{\alpha}\right)}{1+\frac{\pi}{\ln 2}\left(\frac{1}{\beta}+\frac{M_0(\gamma)}{\alpha}\right)}\right|\|L^{-1}\|,$$

where C_1 is a constant independent of f, γ, and β. Consider the function

$$F(\beta)=\frac{\left(\frac{1}{\beta}+\frac{M_0(\gamma)}{\alpha}\right)}{1+\frac{\pi}{\ln 2}\left(\frac{1}{\beta}+\frac{M_0(\gamma)}{\alpha}\right)}.$$

For small β,

$$\left|F(\beta) - \frac{\ln 2}{\alpha\pi}\left(1 - \beta\frac{\ln 2}{\pi} + \beta^2\frac{\ln 2}{\pi}\left(\frac{M_0(\gamma)}{\alpha} + \frac{\ln 2}{\pi}\right)\right)\right| < C_2\beta^3, \quad C_2 = \text{const.}$$

Then, in a certain interval $(0, \beta_0)$,

$$|F(\beta)| \leq \left|\frac{\ln 2}{\alpha\pi}\right| + \beta_0\left|\frac{\ln^2 2}{\alpha\pi^2}\right| + \beta_0^2\left|\frac{\ln^2 2}{\alpha\pi^2}\left(\frac{M_0(\gamma)}{\alpha} + \frac{\ln 2}{\pi}\right)\right| + C_2\beta_0^3.$$

Thus, for any γ, function $F(\beta)$ is uniformly bounded in the interval $(0, \beta_0)$, which proves the required estimate. □

Theorem 19 *For sufficiently small l, the resolvent set of operator-valued function $\tilde{K}(\gamma, l)$ is not empty.*

□ Choose a regular point γ of operator-valued function $N(\gamma, l)$. Since singular points do not depend on l, γ is a regular point for all l. The norm of $N(\gamma, l)$ is bounded uniformly with respect to l, and, according to Lemma 5, the norm of $L_1^{-1}(\gamma, \beta)$ is uniformly bounded in a certain interval $(0, \beta_0)$. Then, for sufficiently small l, $|l^2 \ln l| \|L_1^{-1}(\gamma, \beta)N(\gamma, l)\| < 1$ and operator $\tilde{K}(\gamma, l)$ is invertible. □

Theorem 20 *For sufficiently small l, the spectrum of operator-valued function $\tilde{K}(\gamma, l)$ is at most countable and consists of isolated CNs.*

□ As follows from Theorem 19, for sufficiently small l, operator $K(\gamma)$ is invertible at least at one point γ. Then, according to [35], this operator is invertible in the whole domain of γ with a possible exception of isolated points. □

Theorem 21 *At any regular point γ of operator-valued function $\tilde{K}(\gamma, l)$, the inverse operator-valued function $\tilde{K}^{-1}(\gamma, l)$ admits the representation in the form of the Neumann series:*

$$\tilde{K}(\gamma, l) = \sum_{n=0}^{\infty}(-1)^n l^{2n} \ln^n l (L_1^{-1}N)^n L_1^{-1}, \tag{1.132}$$

which converges in the operator norm uniformly with respect to γ and l in the vicinity of γ and in a certain interval $(0, l_0)$, respectively.

□ At any regular points γ of $\tilde{K}(\gamma, l)$,

$$\tilde{K}^{-1}(\gamma, l) = \left(L_1(\gamma, \beta) + l^2 \ln l N(\gamma, l)\right)^{-1} = (I + l^2 \ln l L_1^{-1}N)^{-1} L_1^{-1}.$$

As follows from Lemma 5, for sufficiently small l, $\|l^2 \ln l L_1^{-1}(\gamma, l)N(\gamma, l)\| < 1$, and operator $(I + l^2 \ln l L_1^{-1}N)^{-1}$ can be expanded in converging Neumann series. Therefore, expansion (1.132) is valid. □

In applications, we will use only the first two terms of expansion (1.132):

$$\tilde{K}^{-1}(\gamma, l)f = L_1^{-1} - l^2 \ln l L_1^{-1} N L_1^{-1} f + O(l^4 \ln^2 l) \quad \text{as} \quad l \to 0. \tag{1.133}$$

In many cases, the following simplified formula can be applied:

$$\tilde{K}^{-1}(\gamma, l)f = L_1^{-1}(\gamma, \beta) + O(l^2 \ln l)$$

$$= \frac{1}{\alpha}L^{-1}f - \frac{1}{\alpha}\frac{\left(\frac{1}{\beta} + \frac{M_0(\gamma)}{\alpha}\right)(L^{-1}f, 1)}{1 + \frac{\pi}{\ln 2}\left(\frac{1}{\beta} + \frac{M_0(\gamma)}{\alpha}\right)}L^{-1}1 + O(l^2 \ln l), \quad \text{as} \quad l \to 0. \tag{1.134}$$

If, in addition, the value of β is sufficiently small, then the following expansion in powers of β can be used:

$$\tilde{K}^{-1}(\gamma,l)f$$
$$= \frac{1}{\alpha}L^{-1}f - \frac{\ln 2}{\alpha\pi}\Big(1 - \beta\frac{\ln 2}{\pi} + \beta^2\frac{\ln 2}{\pi}\Big(\frac{M_0(\gamma)}{\alpha} + \frac{\ln 2}{\pi}\Big)\Big)(L^{-1}f,1)L^{-1}1 \tag{1.135}$$
$$+O(\beta^3) \text{ as } \beta \to 0.$$

For fuctions f satisfying the condition $(L^{-1}f,1) = 0$ (e.g., for odd functions f), the formula expressing $\tilde{K}^{-1}(\gamma,1)f$ is much simpler:

$$\tilde{K}^{-1}(\gamma,l)f = \frac{1}{\alpha}L^{-1}f + O(l^2\ln l), \quad \text{as} \quad l \to 0. \tag{1.136}$$

Formulas (1.133)–(1.136) can be used for the approximate inversion of operator-valued function $\tilde{K}^{-1}(\gamma,l)$ when γ is sufficiently distant from singular points of $K(\gamma)$. Note however that according to Theorem 15, for small l, there exists at least one CN in a neighbourhood of a given pole, i.e., the point, at which the invertibility of $\tilde{K}(\gamma,l)$ is violated. Below, we will take into account the closeness to a simple first-order pole and modify the inversion formulas for this case.

1.10.2 Approximate semi-inversion

Assume, that in the vicinity of a chosen pole γ_{NM}, operator-valued function $\tilde{K}(\gamma,l)$ can be represented as

$$\tilde{K}(\gamma,l)\tilde{\varphi} = K_{NM}(\gamma,l)\tilde{\varphi} + \frac{\mu_{NM}}{\gamma-\gamma_{NM}}(\tilde{\varphi},\varphi_N)\varphi_N,$$

where $K_{NM}(\gamma,l)$ is an operator-valued function holomorphic in the vicinity of γ_{NM}, μ_{NM} is a constant, and $\varphi_N = \varphi_N(lt)$ is a smooth function. In some important cases, $\varphi_N = \cos(lNt/c)$, $\sin(lNt/c)$, or $\exp(ilNt)$.

In fact,

$$K_{NM}(\gamma,l) = L_1(\gamma,\beta) + l^2\ln l N_{NM}(\gamma,l),$$

where $N_{NM}(\gamma,l)$ coincides with $N(\gamma,l)$ minus the singularity at γ_{NM}. Therefore, the formulas for inversion of $\tilde{K}(\gamma,l)$ are applicable to $K_{NM}(\gamma,l)$ with the substitution $M_0(\gamma) \to C_{NM}(\gamma)$, where

$$C_{NM}(\gamma) = M_0(\gamma) - \frac{\mu_{NM}\varphi_N(0)}{\gamma-\gamma_{NM}}. \tag{1.137}$$

Thus, in the vicinity of γ_{NM},

$$K_{NM}^{-1}(\gamma,l)f = L_1^{-1} - l^2\ln l L_1^{-1}N_{NM}L_1^{-1}f + O(l^4\ln^2 l), \quad \text{as} \quad l \to 0.$$

Consider the local representation of equation (1.130) in the form

$$K_{NM}(\gamma,l)\tilde{\varphi} + \frac{\mu_{NM}}{\gamma-\gamma_{NM}}(\tilde{\varphi},\varphi_N)\varphi_N = \tilde{f}. \tag{1.138}$$

Applying operator $K_{NM}^{-1}(\gamma, l)$ to both sides of equation (1.138), we obtain the equivalent equation

$$\tilde{\varphi} + \frac{\mu_{NM}}{\gamma - \gamma_{NM}}(\tilde{\varphi}, \varphi_N) K_{NM}^{-1}(\gamma, l)\tilde{\varphi}_N = K_{NM}^{-1}(\gamma, l)\tilde{f}. \tag{1.139}$$

The solution to (1.139) is uniquely defined if $(\tilde{\varphi}, \varphi_N)$ is uniquely defined. Calculating the scalar product of both sides of equation (1.139) with φ_N, we obtain

$$(\tilde{\varphi}, \varphi_N) = \frac{(\gamma - \gamma_{NM})(K_{NM}^{-1}\tilde{f}, \varphi_N)}{\gamma - \gamma_{NM} + \mu_{NM}(K_{NM}^{-1}\varphi_N, \varphi_N)}. \tag{1.140}$$

Substituting expression (1.140) into formula (1.139), we obtain the local representation of inverse operator-valued function $\tilde{K}^{-1}(\gamma, l)$ in the vicinity of pole γ_{NM}:

$$\tilde{\varphi} = K_{NM}^{-1}(\gamma, l)\tilde{f} - \frac{\mu_{NM}(K_{NM}^{-1}(\gamma, l)\tilde{f}, \varphi_N)}{\gamma - \gamma_{NM} + \mu_{NM}(K_{NM}^{-1}(\gamma, l)\varphi_N, \varphi_N)} K_{NM}^{-1}(\gamma, l)\tilde{\varphi}_N. \tag{1.141}$$

Quantity $(\tilde{\varphi}, \varphi_N)$ is uniqely defined for any $\tilde{f}$ if and only if the denominator of the fraction in (1.141) is not equal to zero. The zeros of the denominator are the points, at which the invertibility of $\tilde{K}(\gamma, l)$ is lost. Since holomorphic operator-valued function $\tilde{K}(\gamma, l)$ is of the Fredholm type, such points are its CNs. Thus, CNs of $\tilde{K}(\gamma, l)$ are the roots of the transcendental equation

$$\gamma = \gamma_{NM} - \mu_{NM}(K_{NM}^{-1}(\gamma, l)\varphi_N, \varphi_N). \tag{1.142}$$

Lemma 6 *For sufficiently small l, there exists a solution to equation (1.142) in a close vicinity of pole γ_{NM}.*

□ Since φ_N is assumed to be a smooth function, we have

$$\varphi_N(lt) = c_N + O(l), \quad \text{as} \quad l \to 0,$$

where c_N is a constant (the case $c_N = 0$ is also possible). Then, according to (1.134), the following approximate representation is valid:

$$(K_{NM}^{-1}(\gamma, l)\varphi_N, \varphi_N) = |c_N|^2 (L_1^{-1}(\gamma, \beta)1, 1) + O(l) \quad \text{as} \quad l \to 0.$$

Assume that l is so small that β is also small. Then, according to formula (1.131),

$$L_1^{-1}(\gamma, \beta)1 = \frac{1}{\alpha}L^{-1}1 - \frac{\ln 2}{\alpha\pi}\left(1 - \beta\frac{\ln 2}{\pi}\right)(L^{-1}1, 1)L^{-1}1 + O(\beta^2), \quad \text{as} \quad \beta \to 0.$$

The values $L^{-1}1$ and $(L^{-1}1, 1)$ can be determined explicitly:

$$L^{-1}1 = \frac{1}{\ln 2}\frac{1}{\sqrt{1 - t^2}}, \qquad (L^{-1}1, 1) = \frac{\pi}{\ln 2}.$$

Then,

$$(L_1^{-1}(\gamma, \beta)1, 1) = \frac{\beta}{\alpha} + O(\beta^2) \quad \text{as} \quad \beta \to 0,$$

and, therefore,

$$(K_{NM}^{-1}(\gamma, l)\varphi_N, \varphi_N) = \beta\frac{|c_N|^2}{\alpha} + O(\beta^2) \quad \text{as} \quad \beta \to 0.$$

Thus, the right-hand side of (1.142) is a contraction mapping, and hence, in the vicinity of γ_{NM}, there exists a unique solution γ^*_{NM} to equation (1.142), which can be obtained as an expansion in powers of β:

$$\gamma^*_{NM} = \gamma_{NM} - \beta\frac{\mu_{NM}|c^2_N|}{\alpha} + \beta^2\frac{\mu_{NM}|c^2_N|}{\alpha}\Big(\frac{\ln 2}{\pi} + \frac{C_{NM}(\gamma_{NM},l)}{\alpha}\Big) + O(\beta^3), \quad \text{as} \quad \beta \to 0. \tag{1.143}$$

In the case when $c_N = 0$,

$$(K^{-1}_{NM}(\gamma,l)\varphi_N,\varphi_N) = O(l) \quad \text{as} \quad l \to 0,$$

and the solution to (1.142) can be obtained in the form

$$\gamma^*_{NM} = \gamma_{NM} + O(l), \quad \text{as} \quad l \to 0. \ \square \tag{1.144}$$

Since the roots of equation (1.142) are CNs of operator-valued function $\tilde{K}(\gamma,l)$, we have proved the following statement.

Theorem 22 *For sufficiently small l, there exists a characteistic number of operator-valued function $\tilde{K}(\gamma,l)$ which lies in the vicinity of pole γ_{NM} and is expressed by formulas (1.143) or (1.144).*

In addition to estimate (1.144), we will derive a more accurate formula for the CN when $c_N = 0$.

Assume that

$$\varphi_N(lt) = lb_N t + O(l^2) \quad \text{as} \quad l \to 0,$$

where $b_N = \text{const}$. Then,

$$(K^{-1}_{NM}(\gamma,l)\varphi_N,\varphi_N) = O(l^2) \quad \text{as} \quad l \to 0,$$

and the solution to equation (1.142) can be obtained as an expansion in powers of l:

$$\gamma^*_{NM} = \gamma_{NM} - \mu_{NM}(K^{-1}_{NM}(\gamma_{NM},l)\varphi_N,\varphi_N)$$

$$+\mu^2_{NM}(K^{-1}_{NM}(\gamma_{NM},l)\varphi_N,\varphi_N)\left(\frac{\partial}{\partial\gamma}\bigg|_{\gamma=\gamma_{NM}}(K^{-1}_{NM}(\gamma,l)\varphi_N,\varphi_N)\right) \tag{1.145}$$

$$+O(l^6) \text{ as } l \to 0.$$

This representation is rather universal; however, it contains the quantity $(K^{-1}_{NM}(\gamma_{NM},l)\varphi_N,\varphi_N)$, which must be calculated. Restrictining ourselves to the error of an order of $O(l^3)$, we will simplify formula (1.145).

Using representation (1.136), we obtain

$$(K^{-1}_{NM}(\gamma,l)\varphi_N,\varphi_N) = l^2\frac{|b_N|^2}{\alpha}(L^{-1}t,t) + O(l^3), \quad \text{as} \quad l \to 0.$$

The values $L^{-1}t$ and $(L^{-1}t,t)$ are determined explicitly:

$$L^{-1}t = \frac{t}{\sqrt{1-t^2}}, \qquad (L^{-1}t,t) = \frac{\pi}{2}.$$

Then, formula (1.145) can be reduced to

$$\gamma^*_{NM} = \gamma_{NM} - l^2 \frac{\pi \mu_{NM} |b_N|^2}{2\alpha} + O(l^3) \text{ as } l \to 0.$$

However, often, more accurate representations of the solution are required. They can be obtained in some special cases, e.g., when φ_N is represented by a system of exponential or trigonometric functions. Consider several important examples.

Example 1 For $\varphi_N = \cos \dfrac{\pi N l}{b} t$, we have

$$K^{-1}_{NM}(\gamma, l)\varphi_N = K^{-1}_{NM}(\gamma, l)1 + O(l^2) = L_1^{-1}(\gamma, \beta)1 + O(l^2)$$

$$= \frac{1}{\alpha \ln 2} \frac{1}{1 + \dfrac{\pi}{\ln 2}\left(\dfrac{1}{\beta} + \dfrac{C_{NM}(\gamma)}{\alpha}\right)} \frac{1}{\sqrt{1-t^2}} + O(l^4 \ln l) \quad \text{as} \quad l \to 0.$$

For sufficiently small l,

$$(K^{-1}_{NM}(\gamma, l)\varphi_N, \varphi_N) = \frac{1}{\alpha}\beta\left(1 - \beta\left(\frac{\ln 2}{\pi} + \frac{C_{NM}(\gamma)}{\alpha}\right)\right) + O(\beta^3) \quad \text{as} \quad \beta \to 0,$$

and the corresponding solution to equation (1.140) is

$$\gamma^*_{NM} = \gamma_{NM} - \beta\frac{\mu_{NM}}{\alpha} + \beta^2 \frac{\mu_{NM}}{\alpha}\left(\frac{\ln 2}{\pi} + \frac{C_{NM}(\gamma_{NM})}{\alpha}\right) + O(\beta^3), \quad \text{as} \quad \beta \to 0.$$

Example 2 For $\varphi_N = \sin \dfrac{\pi N l}{b} t$,

$$\varphi_N = l\frac{\pi N}{b} t + O(l^3).$$

As follows from formulas (1.133),

$$K^{-1}_{NM}(\gamma, l)\varphi_N = l\frac{\pi N}{\alpha b} L^{-1}t - l^3 \ln l \frac{\pi N}{\alpha b} L_1^{-1}(\gamma, \beta) N(\gamma, l) L^{-1}t + O(l^3) \quad \text{as} \quad l \to 0.$$

Then,

$$(K^{-1}_{NM}(\gamma, l)\varphi_N, \varphi_N) = l^2 \frac{\pi^2 N^2}{\alpha b^2}(L^{-1}t, t) - l^4 \ln l \frac{\pi^2 N^2}{\alpha b^2}(L_1^{-1}(\gamma, \beta) N(\gamma, l) L^{-1}t, t) + O(l^4)$$

$$= l^2 \frac{\pi^3 N^2}{2\alpha b^2} - l^4 \ln l \frac{\pi^2 N^2}{\alpha b^2}(L_1^{-1}(\gamma, \beta) N(\gamma, l) \frac{t}{\sqrt{1-t^2}}, t) + O(a^4) \quad \text{as} \quad l \to 0.$$

The value $(L_1^{-1}(\gamma, \beta) N(\gamma, l) \dfrac{t}{\sqrt{1-t^2}}, t)$ can be calculated explicitly for certain operator-valued functions $N(\gamma, l)$. For example, when

$$N(\gamma, l) = \nu(\gamma, l) P_2, \tag{1.146}$$

where $\nu(\gamma, l)$ is a function and P_2 is the projector defined as

$$P_2 \varphi = \int_{-1}^{1} (t_0 - t)^2 \varphi(t) dt,$$

we obtain

$$P_2L^{-1}t = -\pi t,$$

$$L_1^{-1}P_2L^{-1}t = \frac{1}{\alpha}L^{-1}P_2L^{-1}t = -\frac{\pi}{\alpha}\frac{t}{\sqrt{1-t^2}},$$

and, therefore,

$$(L_1^{-1}P_2L^{-1}t, t) = -\frac{\pi^2}{2\alpha}. \tag{1.147}$$

However, in general, the solution to equation (1.142) has the form

$$\gamma_{NM}^{*} = \gamma_{NM} - l^2\mu_N\frac{\pi^3N^2}{2\alpha b^2} + O(l^4\ln l), \quad \text{as} \quad l \to 0. \tag{1.148}$$

Example 3 For $\varphi_N = \exp(ilNt)$, one can use the formula $\varphi_N = \cos lNt + i\sin lNt$ to reduce this case to the previous two.

Chapter 2

Generalized potentials with logarithmic kernels

In the theory of boundary value problems, the integrals

$$u(\mathbf{x}) = \int_C E(\mathbf{x},\mathbf{y})\xi(\mathbf{y})dl_y, \quad v(\mathbf{x}) = \int_C \frac{\partial}{\partial \mathbf{n}_y} E(\mathbf{x},\mathbf{y})\eta(\mathbf{y})dl_y$$

are called the potentials. Here, $\mathbf{x} = (x, y)$, $\mathbf{y} = (x_0, y_0) \in \mathrm{R}^2$; $E(\mathbf{x},\mathbf{y})$ is the fundamental solution of a second-order elliptic differential operator; $\dfrac{\partial}{\partial \mathbf{n}_y}$ is the normal derivative at point $\mathbf{y}$ of the closed piecewise smooth boundary C of a domain in R^2; and $\xi(\mathbf{y})$ and $\eta(\mathbf{y})$ are sufficiently smooth functions defined on C. In the case of the Helmholtz operator $\mathcal{H}(k) = \Delta + k^2$, one can take $E(\mathbf{x},\mathbf{y})$ in the form

$$E(\mathbf{x},\mathbf{y}) = \mathcal{E}(\mathbf{x}-\mathbf{y}) = \frac{i}{4} H_0^{(1)}(k|\mathbf{x}-\mathbf{y}|),$$

where $H_0^{(1)}(z) = -2ig(z) + h(z)$ is the Hankel function, $\dfrac{1}{2} g(\mathbf{x}-\mathbf{y}) = \dfrac{1}{2\pi} \ln \dfrac{1}{|\mathbf{x}-\mathbf{y}|}$ is the kernel of the two-dimensional single layer potential, and the second derivative of $h(z)$ has a logarithmic singularity.

In this chapter, we will consider some properties of the generalized potentials and the Green's potentials in the two-dimensional domains of a certain class together with application of the method of generalized potentials to solving boundary value problems that are reduced to boundary logarithmic integral equations. These problems arise, in particular, in the mathematical models of the propagation and diffraction of electromagnetic and acoustic waves. Typical formulations are given in Chapter 4. We will present several explicit formulas for the kernels of the Green's potentials for canonical domains.

2.1 Definitions of the generalized single layer and double layer potentials

Denote by $S_\Pi = S_\Pi(\Gamma)$ a simple-connected two-dimensional domain bounded by the closed or unclosed piecewise smooth contour ∂S_Π containing a rectilinear part $\Gamma_0 \subset \partial S_\Pi$ such that $\Gamma_0 \subset \{\mathbf{x} : y = 0\}$. We will denote by Γ a subset of Γ_0 assuming that

$\Gamma_0 = \{\mathbf{x} : y = 0,\ x \in [c,d]\}$ and $\Gamma = \{\mathbf{x} : y = 0,\ x \in (a,b)\}$, where $c < a < b < d$; and the endpoints (edges) of Γ have the x-coordinates a and b.

Let us say that functions $U_l(\mathbf{x})$ are the generalized single layer (SLP) ($l = 1$) or double layer (DLP) ($l = 2$) potentials if

$$U_l(\mathbf{x}) = \int_\Gamma K_l(\mathbf{x},t)\varphi_l(t)dt, \quad \mathbf{x} = (x,y) \in S_\Pi(\Gamma),$$

where

$$K_l(\mathbf{x},t) = g_l(\mathbf{x},t) + F_l(\mathbf{x},t) \quad (l = 1,2),$$

$$g_1(\mathbf{x},t) = g(x,y^0) = \frac{1}{\pi}\ln\frac{1}{|\mathbf{x}-\mathbf{y}^0|}, \quad g_2(\mathbf{x},t) = \frac{\partial}{\partial y_0}g(\mathbf{x},\mathbf{y}^0) \quad [\mathbf{y}^0 = (t,0)],$$

$F_{1,2}$ are smooth functions, and we shall assume that for every closed domain $S_{0\Pi}(\Gamma) \subset S_\Pi(\Gamma)$, the following conditions hold

i) $F_1(\mathbf{x},t)$ is once continuously differentiable with respect to the variables of $\mathbf{x}$ and continuous in t;

ii) $F_2(\mathbf{x},t)$ and

$$F_2^1(\mathbf{x},t) = \frac{\partial}{\partial y}\int_q^t F_2(x,s)ds, \quad q \in \mathrm{R}^1,$$

are continuous.

We shall also assume that the densities of the generalized potentials $\varphi_1 \in X_a(\Gamma)$ and $\varphi_2 \in X_0(\Gamma)$, where functional spaces X_a and X_0 are defined in Section 1.1.

The definition of the generalized SLP and DLP is thus connected with the domains from a certain class $S_\Pi(\Gamma)$ with the boundaries containing several rectilinear intervals (Γ).

The following statements are known for the logarithmic potentials. They can be proved straightforwardly for the generalized potentials considered in the S_Π-domains.

Lemma 7

$$\frac{\partial}{\partial y}\int_\Gamma g_2(\mathbf{x},t)\varphi_2(t)dt = \int_\Gamma \frac{\partial}{\partial x}g_1(\mathbf{x},t)\frac{d\varphi_2}{dt}(t)dt,$$

$$x \in S_\Pi, \qquad \varphi_2(t) \in X_0(\Gamma).$$

Lemma 8

$$\lim_{\mathbf{x}\to\mathbf{x}^0\in\Gamma}\frac{\partial}{\partial y}\int_\Gamma g_2(\mathbf{x},t)\varphi_2(t)dt = \int_\Gamma \lim_{x\to x^0\in\Gamma}\frac{\partial}{\partial x}g_1(\mathbf{x},t)\frac{d\varphi_2}{dt}dt = S\frac{d\varphi_2}{dt};$$

$$Sf(t) \equiv \frac{1}{\pi}\int_\Gamma \frac{f(t')}{t'-t}dt', \quad \varphi_2(t) \in X_0(\Gamma).$$

Lemma 9 *The properties of the normal and tangential derivatives of SLP and of the trace of DLP on L coincide with those of the logarithmic potentials.*

Lemma 10 *There exists the limiting value (trace) of the normal derivative of DLP in S_Π-domain and it admits the representation as a singular integral:*

$$\lim_{\mathbf{x}\to\mathbf{x}^0\in\Gamma}\frac{\partial U_2}{\partial y}=\lim_{\mathbf{x}\to\mathbf{x}^0\in\Gamma}\int_\Gamma\left[\frac{\partial g_1}{\partial x}+F_2^1\right](\mathbf{x},t)\frac{d\varphi_2}{dt}dt=(S+K_2^1)\frac{d\varphi_2}{dt},$$

where

$$\varphi_2(t)\in X_0(\Gamma),\quad K_2^1 f=\int_\Gamma F_2^1(\mathbf{x},t)f(t)dt.$$

The proof of Lemma 10 is based on the classical results of the two-dimensional potential theory and is obtained from the previous lemmas and the definition of DLP under the condition that its density belongs to $X_0(\Gamma)$. The behaviour of the limiting value of the DLP normal derivative near the edges of Γ requires a special consideration. We just note here that by applying the results of [1], one can show that

$$\left|\frac{\partial U_2}{\partial x_m}\right|<\rho^\alpha(\mathbf{x},\mathbf{x}_\Gamma),\quad m=1,2;$$

$$\eta(s)\in X_0(\Gamma),\quad \eta(s)=\lim_{\mathbf{x}\to\mathbf{x}^0\in\Gamma}\frac{\partial U_2}{\partial y},\quad \mathbf{x}^0=(s,0),$$

where $\rho(\mathbf{x},\mathbf{x}_\Gamma)$ is the distance from $\mathbf{x}$ to one of the edges $\mathbf{x}_\Gamma$ of Γ and $\alpha>-1$. Details of the proofs of these lemmas are given in [38].

All statements of Lemmas 7–10 are valid for a more general family of the domains $S_\Pi^c(\Gamma)$ bounded by a closed piecewise smooth contours ∂S_Π^c such that

$$\Gamma\subset\Gamma_0\subset\partial S_\Pi^c,\quad \Gamma_0\subset\bigcup\{y=\text{const}\},$$

$S_\Pi\subset S_\Pi^c$ and Γ_0 and $\Gamma\subset\Gamma_0$ consist of a finite number of nonintersecting finite rectilinear intervals:

$$\Gamma_0=\bigcup_{l=1}^{N_L}\{\mathbf{x}:\,y=q_l,\,x\in[a_{2l-1},a_{2l}]\},\quad \Gamma=\bigcup_{l=1}^{N_\Gamma}\Gamma^{(l)},$$

$$\Gamma^{(l)}=\{\mathbf{x}:\,y=q_l,\,x\in(d_{2l-1},d_{2l})\},$$

$$a_{2l-1}<d_{2l-1}<d_{2l}<a_{2l},\quad l=1,2,\ldots,N_\Gamma.$$

The endpoints of Γ $\mathbf{x}_{2l-1}=(d_{2l-1},q_l)$ and $\mathbf{x}_{2l}=(d_{2l},q_l)$, $l=1,2,\ldots N_\Gamma$ are called the edges. In fact, here, it is sufficient to reformulate the proofs obtained in the case $N_\Gamma=1$ (when $\Gamma=\Gamma^{(1)}$ consists of one interval on the axis $y=0$) applying the local Cartesian coordinate system connected with every rectilinear component $\Gamma^{(1)}$ of Γ and considering the densities of the potentials as restrictions on this rectilinear component [and as the elements of spaces $X_a(\Gamma^{(l)})$ and $X_0(\Gamma^{(l)})$].

2.2 Green's potentials

Let us introduce the Green's functions $G^i=G^i(\mathbf{x},\mathbf{y};k)$ of the ith boundary-value problem ($i=1,2$) for the Helmholtz equation $\mathcal{H}(k)U=0$ of a (bounded or unbounded) domain $\Omega\in S_\Pi^c$ (under the assumption of its existence). In particular,

$$h^l(\mathbf{x},\mathbf{y};k)=\mathcal{E}(\mathbf{x}-\mathbf{y})+(-1)^l\mathcal{E}(\mathbf{x}-\mathbf{y}^*),$$

$$\mathbf{y} = (x_0, y_0), \quad \mathbf{y}^* = (x_0, -y_0),$$

which is the Green's function of the first ($l = 1$) or the second ($l = 2$) boundary value problem for the Helmholtz equation of the half-plane $\Omega^+ = \{\mathbf{x} : y > 0\}$, is the kernel of SLP for $l = 2$. The function

$$\frac{\partial}{\partial y_0} h^1(\mathbf{x}, \mathbf{y}^0; k)$$

is the kernel of DLP for $l = 1$ in every $S^c_\Pi(\Gamma)$-domain such that $S^c_\Pi(\Gamma) \subset \{\mathbf{x} : y > 0\}$.

We will call the integrals

$$u(\mathbf{x}) = \int_\Gamma G^2(\mathbf{x}, \mathbf{y}^0; k)\varphi(t)dt, \quad x \in \Omega, \quad \mathbf{y}^0 = (t, 0) \in \Gamma, \qquad \varphi(t) =\in \widetilde{H_\mu}(\Gamma); \tag{2.1}$$

$$v(\mathbf{x}) = \int_\Gamma K_1(\mathbf{x}, \mathbf{y}^0; k)\psi(t)dt, \quad \mathbf{x} \in \Omega, \quad \mathbf{y}^0 = (t, 0) \in \Gamma, \qquad K_l(\mathbf{x}, \mathbf{y}^0; k) = \frac{\partial}{\partial y_0} G^1(\mathbf{x}, \mathbf{y}^0; k), \quad \psi(t) =\in X_0(\Gamma) \tag{2.2}$$

the single layer and double layer Green's potentials.

Theorem 23 ([38]) *The single layer and double layer Green's potentials (2.1) and (2.2) are the generalized SLP and DLP, respectively, in the $S^c_\Pi(L)$-domains. The functions*

$$G^2(\mathbf{x}, \mathbf{y}^0; k) \quad \text{and} \quad \frac{\partial}{\partial y_0} G^1(\mathbf{x}, \mathbf{y}^0; k)$$

admit representations as the kernels of SLP and DLP.

□ The proof for a (bounded) domain $S_\Pi(\Gamma)$ is based on the study of the integral representations for the solutions of two auxiliary problems

$$(\Delta + k^2)u = 0, \quad (\Delta + k^2)v = 0, \quad x \in S_\Pi(\Gamma),$$

$$\left.\frac{\partial u}{\partial n}\right|_\Gamma = \varphi \in X_a(\Gamma) \quad v|_\Gamma = \psi \in X_0(\Gamma),$$

where it is assumed that $\operatorname{supp}\varphi \subset L$ and $\operatorname{supp}\psi \subset L$.

To derive the required representations, it is sufficient to apply the Green's identity to functions $u(\mathbf{x})$ and $V_2(\mathbf{x}, \mathbf{y}; k) = h_2(\mathbf{x}, \mathbf{y}; k) - G^2(\mathbf{x}, \mathbf{y}; k)$ and $v(\mathbf{x})$ and $V_1(\mathbf{x}, \mathbf{y}; k) = h_1(\mathbf{x}, \mathbf{y}; k) - G^1(\mathbf{x}, \mathbf{y}; k)$, and to directly verify the fulfillment of conditions (i) and (ii) formulated in the definition of generalized potentials. It is also taken into account that h_2 and $\frac{\partial}{\partial y_0} h_1$ are the kernels of SLP and DLP, respectively.

The proof for a domain $S^c_\Pi(\Gamma)$ is based on the separate consideration of integral representations in a vicinity of every rectilinear component of Γ (coinciding with a part of the line $y = c \neq 0$) using the Green's functions $h^{1,2}_c$ of the half-plane $\Omega^{+c} = \{\mathbf{x} : y > c\}$.
□

The direct application of Lemma 10 yields [in the case of a bounded $S_\Pi(\Gamma)$-domain] the following

Theorem 24 (**[38]**) *There exist limiting values (traces) on* Γ *of the first (tangential) derivative of the single layer Green's potential* u *given by (2.1) and of the normal derivative of the double layer Green's potential* v *given by (2.2) in the* $S_\Pi(\Gamma)$*-domains; they admit representations as singular integrals:*

$$\lim_{\mathbf{x}\to\mathbf{x}^0\in\Gamma}\frac{\partial u}{\partial x}=\left(S+\hat{G}\right)\varphi=\frac{1}{\pi}\int_\Gamma\frac{\varphi(t)}{t-x}dt+\int_\Gamma\hat{g}(\mathbf{x}^0,t;k)\varphi(t)dt;$$

$$\lim_{\mathbf{x}\to\mathbf{x}^0\in\Gamma}\frac{\partial v}{\partial y}=\lim_{\mathbf{x}\to\mathbf{x}^0\in\Gamma}\int_\Gamma\left(\frac{\partial g_1}{\partial x}+g_2^1\right)(\mathbf{x},t;k)\frac{d\varphi_2}{dt}dt=(S+G_2^1)\frac{d\varphi_2}{dt},$$

where $\varphi_2\in X_0(\Gamma)$ *and* $\varphi\in X_a(\Gamma)$*;*

$$G_2^1f=\int_\Gamma g_2^1(\mathbf{x}^0,t;k)f(t)dt,$$

$$g_2^1(\mathbf{x}^0,t;k)=\frac{\partial}{\partial y}\bigg|_{\mathbf{x}=\mathbf{x}^0\in\Gamma}\int_q^t\left[\frac{\partial}{\partial y_0}G^1(\mathbf{x},\mathbf{y}^0;k)-\frac{\partial}{\partial y_0}h^1(\mathbf{x},\mathbf{y}^0;k)\right]ds,\quad q\in R^1;$$

$\hat{g}(\mathbf{x}^0,t;k)$ *is once continuously differentiable and* $g_2^1(\mathbf{x}^0,t;k)$ *is continuous with respect to* $x,\ t\in\Gamma$.

One can generalize the statement of Theorem 24 for an arbitrary S_Π^c-domain, if to apply, as well as in Lemma L2.4, the local Cartesian coordinate system connected with every rectilinear component $\Gamma^{(1)}$ of Γ and consider the densities of the potentials as restrictions on this rectilinear component [and as the elements of spaces $X_a(\Gamma^{(l)}$ and $X_0(\Gamma^{(l)})$].

We note also important properties of the Green's functions $G^1(\mathbf{x},\mathbf{y};\lambda)$ and $G^2(\mathbf{x},\mathbf{y};\lambda)$ of an $S_\Pi(\Gamma)$-domain of the Helmholtz equation $\Delta+\lambda\varepsilon$ considered with respect to spectral parameter λ; here, ε is a constant. Using these facts, one can construct an analytical continuation of the Green's potentials to the domain of complex (spectral) parameter λ.

The traces of the Green's functions of $S_\Pi(\Gamma)$-domains

$$G^2(\mathbf{x},\mathbf{y};\lambda)\Big|_{\mathbf{y}\in\Gamma},\quad G^2(\mathbf{x},\mathbf{y};\lambda)\Big|_{\mathbf{x},\mathbf{y}\in\Gamma}$$

and

$$\frac{\partial G^1}{\partial y}\bigg|_{\mathbf{y}\in\Gamma}(\mathbf{x},\mathbf{y};\lambda),\quad \frac{\partial G^1}{\partial y}\bigg|_{\mathbf{x},\mathbf{y}\in\Gamma}(\mathbf{x},\mathbf{y};\lambda).$$

admit analytical continuation as meromorphic functions of λ on the whole complex plane C_λ with real finite-order poles, so that, for example,

$$G^2(\mathbf{x}^0,\mathbf{y}^0;\lambda)=\frac{u_n^{(2)}(\lambda;t)u_n^{(2)}(\lambda;s)}{\lambda-\lambda_n^{(2)}}+N_2^{2n}(\lambda;t,s);\tag{2.3}$$

$$N_2^{2n}(\lambda;t,s)=c_2\ln|s-t|+N_{21}^{2n}(\lambda;t,s),\quad \mathbf{x}^0=(t,0)\in\Gamma,\ \mathbf{y}^0=(s,0)\in\Gamma,$$

and these representations hold in a vicinity of its every pole $\lambda_n^{(2)}$. Here, $u_n^{(2)}$ are the eigenfunctions of the second boundary value problem in Ω^2 for $\Delta+\lambda\varepsilon$; N_2^{2n} and N_{21}^{2n} are meromorphic functions of λ with the same poles as the Green's function has [except for $\lambda_n^{(2)}$ in the case of N_2^{2n}]; N_2^{2n} is holomorphic in a vicinity of $\lambda_n^{(2)}$; and N_{21}^{2n} is once continuously differentiable with respect to $t,s\in L$ $(n=0,1,\ldots)$.

These properties of the Green's functions enable one to apply the technique of the generalized Green's potentials to solving boundary eigenvalue problems for the Helmholtz equation in $S_\Pi(\Gamma)$-domains. In particular, representation (2.3) makes it possible to reduce such problems to the determination of CNs of boundary integral finite-meromorphic operator-valued functions with a logarithmic singularity of the kernel that admit separation of pole pencils (described in Section 1.7). The details of this approach are set forth in Chapter 4.

2.3 Examples for canonical domains

In this section, we present explicit expressions for the Green's functions of the boundary value problem

$$\Delta u + k^2 u = f,$$
$$\frac{\partial u}{\partial n} = 0,$$

in some canonical domains. When necessary, we will use the notation $\mathbf{x} = (x, y)$ and $\mathbf{y} = (x_0, y_0)$ in Cartesian coordinates or $\mathbf{x} = (r, \phi)$ and $\mathbf{y} = (r_0, \phi_0)$ in polar coordinates. However, for the sake of convenient applications, all explicit representations will be written using variables x, y, x_0, and y_0 and r, ϕ, r_0, and ϕ_0 .

2.3.1 Half-plane

For the half-plane $\{(x, y) : -\infty < x < \infty, 0 < y < \infty\}$, the Green's function has the form

$$G(x_0, y_0, x, y; k) = \frac{i}{4} H_0^{(1)}(k\sqrt{(x_0 - x)^2 + (y_0 - y)^2}) + \frac{i}{4} H_0^{(1)}(k\sqrt{(x_0 - x)^2 + (y_0 + y)^2}).$$

At the boundary points $(x, 0)$ and $(x_0, 0)$,

$$G(x_0, 0, x, 0; k) = \frac{i}{2} H_0^{(1)}(k|x_0 - x|).$$

Using the known representation of the Hankel function [77], we can separate the logarithmic singularity:

$$\begin{aligned} G(x_0, 0, x, 0; k) = {} & \frac{1}{\pi} \ln \frac{1}{|x_0 - x|} + \frac{1}{\pi} \ln \frac{2}{\gamma k} - \frac{1}{\pi} \sum_{n=1}^{\infty} \frac{(-1)^n (k|x_0 - x|)^{2n}}{(n!)^2 4^n} \ln \frac{\gamma k |x_0 - x|}{2} \\ & + \frac{1}{2\pi} \sum_{n=0}^{\infty} \frac{(-1)^n (k|x_0 - x|)^{2n}}{(n!)^2 4^n} \sum_{m=1}^{n} \frac{1}{m} + \frac{i}{2} \sum_{n=0}^{\infty} \frac{(-1)^n (k|x_0 - x|)^{2n}}{(n!)^2 4^n}. \end{aligned} \tag{2.4}$$

where $\gamma = 1.781072418\ldots$ is the Euler constant [77].

As a function of k, $G(x_0, y_0, x, y; k)$ is analytical in the whole complex plane for all $\mathbf{x} \neq \mathbf{y}$ and has only one singular point, namely, a logarithmic branch point at $k = 0$.

2.3.2 Rectangle

For the rectangle $\{(x, y) : 0 < x < c, 0 < y < b\}$, the Green's function has the form

$$G(x_0, y_0, x, y; k) = \frac{4}{bc} \sum_{n=0}^{\infty} \sum_{m=0}^{\infty} \frac{\delta_n \delta_m \cos(\pi n x_0 / c) \cos(\pi n y_0 / b) \cos(\pi n x / c) \cos(\pi n y / b)}{\lambda_{nm} - k^2},$$

where

$$\delta_n = \begin{cases} 1/2, & n = 0, \\ 1, & n = 1, 2, \ldots \end{cases}$$

and

$$\lambda_{nm} = \pi^2 \left(\frac{n^2}{c^2} + \frac{m^2}{b^2} \right), \quad n, m = 0, 1, 2, \ldots.$$

At the boundary points $(x, 0)$ and $(x_0, 0)$ $(0 < x < c,\ 0 < x_0 < c)$,

$$G(x_0, 0, x, 0; k) = \frac{4}{bc} \sum_{n=0}^{\infty} \left(\sum_{m=0}^{\infty} \frac{\delta_m}{\lambda_{nm} - k^2} \right) \delta_n \cos \frac{\pi n x_0}{c} \cos \frac{\pi n x}{c}.$$

After separating the slowly converging trigonometric series that correspond to the logarithmic singularity, we obtain the following representation:

$$\begin{aligned} G(x_0, 0, x, 0; k) &= \frac{2}{\pi} \sum_{n=1}^{\infty} \frac{\cos(\pi n x_0/c) \cos(\pi n x/c)}{n} \\ &+ \frac{2}{\pi} \sum_{n=1}^{\infty} \frac{e^{-\pi n b/c}}{n \sinh(\pi n b/c)} \cos \frac{\pi n x_0}{c} \cos \frac{\pi n x}{c} + \frac{2}{bc} \sum_{m=0}^{\infty} \frac{\delta_m}{\lambda_{0m} - k^2} \\ &+ \frac{4}{bc} \sum_{n=1}^{\infty} \sum_{m=0}^{\infty} \delta_m \left(\frac{1}{\lambda_{nm} - k^2} - \frac{1}{\lambda_{nm}} \right) \cos \frac{\pi n x_0}{c} \cos \frac{\pi n x}{c}. \end{aligned} \tag{2.5}$$

The first trigonometric series in (2.5) with coeficient $\frac{1}{n}$ is the Fourier expansion of a logarithmic function:

$$\frac{2}{\pi} \sum_{n=1}^{\infty} \frac{\cos(\pi n x_0/c) \cos(\pi n x/c)}{n} = -\frac{1}{\pi} \ln \left| 2 \sin \frac{\pi(x_0 - x)}{2c} \right| - \frac{1}{\pi} \ln \left(2 \sin \frac{\pi(x_0 + x)}{2c} \right),$$

which has the required logarithmic singularity. As functions of x and x_0, the second series in (2.5) is infinitely differentiable, the third is constant, and the fourth is once continuously differentiable.

$G(x_0, y_0, x, y; k)$ is a meromorphic function of k for all $\mathbf{x} \neq \mathbf{y}$ with the countable number of first-order poles at points $\pm\sqrt{\lambda_{nm}}$ $(n, m = 0, 1, 2, \ldots)$.

2.3.3 Circle

For the circle $\{(r, \phi) :\ 0 \le r < R,\ 0 \le \phi < 2\pi\}$, the Green's function has the form

$$G(r_0, \phi_0, r, \phi; k)$$

$$= \begin{cases} \dfrac{1}{4} \displaystyle\sum_{n=-\infty}^{+\infty} \frac{J_n(kr)}{J_n'(kR)} (J_n(kr_0) N_n'(kR) - J_n'(kR) N_n(kr_0)) e^{in(\phi - \phi_0)}, & r < r_0 \le R \\ \dfrac{1}{4} \displaystyle\sum_{n=-\infty}^{+\infty} \frac{J_n(kr_0)}{J_n'(kR)} (J_n(kr) N_n'(kR) - J_n'(kR) N_n(kr)) e^{in(\phi - \phi_0)}, & r_0 < r \le R. \end{cases}$$

At the boundary points (R, ϕ_0) and (R, ϕ),

$$G(R, \phi_0, R, \phi; k) = \frac{1}{\pi k R} \sum_{n=0}^{\infty} \frac{\delta_n J_n(kR)}{J_n'(kR)} \cos n(\phi - \phi_0). \tag{2.6}$$

After separating slowly converging trigonometric series, we obtain the following representation:

$$G(R,\phi_0,R,\phi;k)=\frac{1}{\pi}\sum_{n=1}^{\infty}\frac{\cos n(\phi-\phi_0)}{n}-\frac{1}{2\pi kR}\frac{J_0(kR)}{J_1(kR)}$$
$$+\frac{kR}{\pi}\sum_{n=1}^{\infty}\frac{J_{n+1}(kR)}{J_n(kR)}\frac{\cos n(\phi-\phi_0)}{n\left(n-kR\dfrac{J_{n+1}(kR)}{J_n(kR)}\right)}. \tag{2.7}$$

The first trigonometric series in (2.7) is the Fourier expansion of a logarithmic function:

$$\frac{1}{\pi}\sum_{n=1}^{\infty}\frac{\cos n(\phi-\phi_0)}{n}=-\frac{1}{\pi}\ln\left|2\sin\frac{(\phi-\phi_0)}{2}\right|. \tag{2.8}$$

Using the estimates for the Bessel functions with large indices [78]

$$\frac{J_{n+1}(kR)}{J_n(kR)}=O\left(\frac{1}{n}\right),\quad \text{as}\quad n\to\infty,$$

one can show that the coefficient of the last trigonometric series in (2.7) has the order $O(n^{-3})$. Therefore, this series is once continiously differentiable with respect to ϕ and ϕ_0.

As a function of k, $G(r_0,\phi_0,r,\phi;k)$ is meromorphic for all $\mathbf{x}\neq\mathbf{y}$ with the countable number of first-order poles at the points $k=\dfrac{j'_{n,m}}{R}$ $(n,m=0,1,2,\ldots)$, where $j'_{n,m}$ denotes the mth root of derivative $J'_n(x)$ of the Bessel function.

2.3.4 Exterior of a circle

For the exterior of the circle $\{(r,\phi):\ R<r<\infty,\ 0\le\phi<2\pi\}$, the Green's function has the form

$$G(r_0,\phi_0,r,\phi;k)$$
$$=\begin{cases}\dfrac{1}{4}\displaystyle\sum_{n=-\infty}^{+\infty}\frac{H_n^{(1)}(kr_0)}{H_n^{(1)\prime}(kR)}(J'_n(kR)N_n(kr)-J_n(kr)N'_n(kR))e^{in(\phi-\phi_0)}, & R\le r<r_0,\\ \dfrac{1}{4}\displaystyle\sum_{n=-\infty}^{+\infty}\frac{H_n^{(1)}(kr)}{H_n^{(1)\prime}(kR)}(J'_n(kR)N_n(kr_0)-J_n(kr_0)N'_n(kR))e^{in(\phi-\phi_0)}, & R\le r_0<r.\end{cases} \tag{2.9}$$

At the boundary points (R,ϕ) and (R,ϕ_0), the following representation can be used:

$$G(R,\phi_0,R,\phi;k)=-\frac{1}{\pi kR}\sum_{n=0}^{\infty}\frac{\delta_n H_n^{(1)}(kR)}{H_n^{(1)\prime}(kR)}\cos n(\phi-\phi_0). \tag{2.10}$$

After separating slowly converging trigonometric series, we obtain

$$G(R,\phi_0,R,\phi;k)=\frac{1}{\pi}\sum_{n=1}^{\infty}\frac{\cos n(\phi-\phi_0)}{n}+\frac{1}{2\pi kR}\frac{H_0^{(1)}(kR)}{J_1^{(1)}(kR)}$$
$$+\frac{kR}{\pi}\sum_{n=1}^{\infty}\frac{H_{n-1}^{(1)}(kR)}{H_n^{(1)}(kR)}\frac{\cos n(\phi-\phi_0)}{n\left(n-kR\dfrac{H_{n-1}^{(1)}(kR)}{H_n^{(1)}(kR)}\right)}. \tag{2.11}$$

The first trigonometric series in (2.11) is the Fourier expansion of logarithmic function (2.8).

According to [78],

$$\frac{H^{(1)}_{n-1}(kR)}{H^{(1)}_{n}(kR)} = O\left(\frac{1}{n}\right), \quad \text{as} \quad n \to \infty,$$

and the coefficient of the last trigonometric series in (2.11) has the order $O(n^{-3})$. Therefore, this series is once continiously differentiable with respect to ϕ and ϕ_0.

As a function of k, $G(r_0, \phi_0, r, \phi; k)$ is analytical for all $\mathbf{x} \neq \mathbf{y}$ in the whole complex plane with the only one singular point at zero.

2.3.5 Ring

For the ring $\{(r, \phi) : R_1 < r < R_2, 0 \leq \phi < 2\pi\}$, the Green's function has the form

$$G(r_0, \phi_0, r, \phi; k) = \frac{1}{4} \sum_{n=-\infty}^{+\infty} \frac{e^{in(\phi-\phi_0)}}{J'_n(kR_1)N'_n(kR_2) - J'_n(kR_2)N'_n(kR_1)}$$

$$\times \begin{cases} [J'_n(kR_2)N_n(kr_0) - J_n(kr_0)N'_n(kR_2)][J_n(kr)N'_n(kR_1) - J'_n(kR_1)N_n(kr)], & r < r_0, \\ [J'_n(kR_1)N_n(kr_0) - J_n(kr_0)N'_n(kR_1)][J_n(kr)N'_n(kR_2) - J'_n(kR_2)N_n(kr)], & r > r_0. \end{cases}$$

At the boundary points (R_2, ϕ) and (R_2, ϕ_0), the following representation can be used:

$$G(R, \phi_0, R, \phi; k) = -\frac{1}{\pi k R_2} \sum_{n=0}^{\infty} \delta_n \frac{J_n(kR_2)N'_n(kR_1) - J'_n(kR_1)N_n(kR_2)}{(J'_n(kR_1)N'_n(kR_2) - J'_n(kR_2)N'_n(kR_1))} \cos n(\phi - \phi_0).$$

After separating slowly converging trigonometric series, we obtain

$$G(R, \phi_0, R, \phi; k) = \frac{1}{\pi} \sum_{n=1}^{\infty} \frac{\cos n(\phi - \phi_0)}{n}$$

$$+ \frac{1}{2\pi k R_2} \frac{J_0(kR_2)N_1(kR_1) - J_1(kR_1)N_0(kR_2)}{(J_1(kR_1)N_1(kR_2) - J_1(kR_2)N_1(kR_1))} \tag{2.12}$$

$$-\frac{kR}{\pi} \sum_{n=1}^{\infty} \frac{A_n}{n^2\left(1 + \dfrac{kR_2 A_n}{n}\right)} \cos n(\phi - \phi_0),$$

where

$$A_n = \frac{J'_n(kR_1)N_{n+1}(kR_2) - J_{n+1}(kR_2)N'_n(kR_1)}{(J_n(kR_2)N'_n(kR_1) - J'_n(kR_1)N_n(kR_2))}.$$

Using the estimates for the Bessel functions with large indices, one can show that

$$A_n = O(n^{-1}).$$

Then, the coefficient of the last trigonometric series in (2.12) has the order $O(n^{-3})$, and the series is once continiously differentiable with respect to ϕ and ϕ_0.

As a function of k, $G(r_0, \phi_0, r, \phi; k)$ is meromorphic for all $\mathbf{x} \neq \mathbf{y}$ with the countable number of first-order poles at the points $k = \lambda_{nm}$ $(n, m = 0, 1, 2, \cdots)$, where λ_{nm} denotes the mth root of the equation

$$J'_n(kR_1)N'_n(kR_2) - J'_n(kR_2)N'_n(kR_1) = 0$$

considered with respect to k.

Chapter 3

Summation operators

In this chapter, we will present, following [27, 31, 43, 44], some results of the spectral theory of the so-called summation operators acting in Hilbert spaces of complex sequences. These operators, which are produced by infinite matrices, will be applied to the solution of logarithmic integral equations.

3.1 Matrix representation of an operator in the Hilbert space and the summation operators

Let $\{\varphi_n\}$ be an orthonormal basis in the Hilbert space H; then each element $f \in H$ can be represented in the form of the generalized Fourier series $f = \sum_{n=1}^{\infty} a_n\varphi_n$, where the sequence of complex numbers $\{a_n\}$ is such that the series $\sum_{n=1}^{\infty} |a_n|^2 < \infty$ converges in H. There exists [43] a (linear) one-to-one correspondence between such sequences $\{a_n\}$ and the elements of the Hilbert space.

Following [43, 44, 50] we will use this isomorphism to construct the matrix representation of a completely continuous operator A and a completely continuous operator-valued function $A(\gamma)$ holomorphic with respect to complex variable γ acting in H.

Assume that $g = Af$, $f \in H$, and

$$g_k = g_k(\gamma) = (A(\gamma)f, \varphi_k),$$

i.e., an infinite sequence (vector) of the Fourier coefficients $\{g_n\} = (g_1, g_2, \ldots)$ corresponds to an element $g \in H$ and we can write $g = (g_1, g_2, \ldots)$. Let $f = \sum_{k=1}^{\infty} f_k\varphi_k$; then $f = (f_1, f_2, \ldots)$. Set $f^N = Pf = \sum_{k=1}^{N} f_k\varphi_k$ and $A\varphi_n = \sum_{k=1}^{\infty} a_{kn}\varphi_k$, $n = 1, 2, \ldots$. Then,

$$\lim_{N\to\infty} Af^N = \lim_{N\to\infty} \sum_{n=1}^{N} f_n \sum_{k=1}^{\infty} a_{kn}\varphi_k = Af = \sum_{k=1}^{\infty}\Big(\sum_{n=1}^{\infty} a_{kn}f_n\Big)\varphi_k.$$

On the other side, $Af = \sum_{k=1}^{\infty} g_k\varphi_k$, hence, $g_k = \sum_{n=1}^{\infty} a_{kn}(Af, \varphi_k) = (g, \varphi_k)$. Now we set $A\varphi_k = g_k$ and

$$(g_k, \varphi_j) = (A\varphi_k, \varphi_j) = a_{jk},$$

where a_{jk} are the Fourier coefficients in the expansion of element g_k over the basis $\{\varphi_j\}$. Consequently,

$$A\varphi_k = g_k = \sum_{j=1}^{\infty} a_{jk}\varphi_j,$$

and the series

$$\sum_{j=1}^{\infty} |a_{jk}|^2 = \sum_{j=1}^{\infty} |(g_k, \varphi_j)|^2 < \infty; \quad k = 1, 2, \ldots.$$

Infinite matrix $\{a_{jk}\}_{j,k=1}^{\infty}$ defines operator A; that is, one can reconstruct A by this matrix and an orthonormal basis $\{\varphi_k\}$ of the Hilbert space H: find the unique element Af for each $f \in H$. In order to prove the latter assertion, we will assume that $f = \sum_{k=1}^{\infty} \xi_k \varphi_k$ and $f^N = Pf$. Then, $Af^N = \sum_{k=1}^{\infty} \eta_k^N \varphi_k$, where $\eta_k^N = \sum_{j=1}^{N} a_{kj}\xi_j$. Applying the continuity of operator A, we have

$$C_k = (Af, \varphi_k) = \lim_{N\to\infty} (Af^N, \varphi_k) = \lim_{N\to\infty} \eta_k^N = \sum_{j=1}^{\infty} a_{kj}\xi_j.$$

Hence, for each $f \in H$,

$$Af = \sum_{k=1}^{\infty} C_k \varphi_k, \qquad C_k = \sum_{j=1}^{\infty} a_{kj}\xi_j,$$

and the latter equalities define the matrix representation of the operator A in the basis $\{\varphi_k\}$ of the Hilbert space H.

Definition 8 *Operators and operator-valued functions acting in the Hilbert space H and given by their matrix representations will be called the summation operators and summation operator-valued functions, respectively.*

Following [43], we will give a definition of the resolvent of the operator-valued function $A(\gamma)$ as the resolvent of a canonical operator-valued function $E - A(\gamma)$ of the second kind, where E is the identity operator.

Definition 9 *The resolvent of operator-valued function $A(\gamma)$ is an operator $R(\gamma)$ such that*

$$(E + R(\gamma))(E - A(\gamma)) = E.$$

Theorem 25 ([36], [43], [50]) *Let $A(\gamma)$ be a completely continuous operator-valued function holomorphic with respect to complex variable γ in a domain G. If for a certain $\gamma_0 \in G$, its resolvent $R(\gamma)$ is a bounded operator, then $R(\gamma)$ exists in the whole domain G except for not more than a countable set of points and is a meromorphic function of γ.*

Corollary 2 ([35], [36], [50]) *In every bounded closed subdomain of the domain G, there may exist not more than a countable set of the CNs of operator-valued function $E - A(\gamma)$.*

3.2 Galerkin methods and basis of Chebyshev polynomials

We describe the approximate solution of linear operator equations considering their projections onto finite-dimensional subspaces, which is convenient for practical calculations. Below, it is assumed that all operators are linear and bounded. Then, the general results concerning the substantiation and convergence the Galerkin methods will be applied to solving logarithmic integral equations. In the sequel, we will use definitions, general properties, and convergence of projection methods set forth in [60].

Definition 10 *Let X and Y be the Banach spaces and let $A : X \to Y$ be an injective operator. Let $X_n \subset X$ and $Y_n \subset Y$ be two sequences of subspaces with* $\dim X_n = \dim Y_n = n$ *and $P_n : Y \to Y_n$ be the projection operators. The projection method generated by X_n and P_n approximates the equation*

$$A\varphi = f$$

by its projection

$$P_n A\varphi_n = P_n f.$$

This projection method is called convergent for operator A, if there exists an index N such that for each $f \in \Im A$, the approximating equation $P_n A\varphi = P_n f$ has a unique solution $\varphi_n \in X_n$ for all $n \geq N$ and these solutions converge, $\varphi_n \to \varphi$, $n \to \infty$, to the unique solution φ of $A\varphi = f$.

In terms of operators, the convergence of the projection method means that for all $n \geq N$, the finite-dimensional operators $A_n := P_n A : X_n \to Y_n$ are invertible and the pointwise convergence $A_n^{-1} P_n A\varphi \to \varphi$, $n \to \infty$, holds for all $\varphi \in X$. In the general case, one may expect that the convergence occurs if subspaces X_n form a dense set:

$$\inf_{\psi \in X_n} \|\psi - \varphi\| \to 0, \quad n \to \infty \tag{3.1}$$

for all $\varphi \in X$. This property is called *the approximating property* (any element $\varphi \in X$ can be approximated by elements $\psi \in X_n$ with an arbitrary accuracy in the norm of X). Therefore, in the subsequent analysis, we will always assume that this condition is fulfilled.

Since $A_n = P_n A$ is a linear operator acting in finite-dimensional spaces, the projection method reduces to solving a finite-dimensional linear system. Below, the Galerkin method will be considered as the projection method under the assumption that the above definition is valid in terms of the orthogonal projection.

Consider first the general convergence and error analysis [60].

Lemma 11 *The projection method converges if and only if there exists an index N and a positive constant M such that for all $n \geq N$, the finite-dimensional operators $A_n = P_n A : X_n \to Y_n$ are invertible and operators $A_n^{-1} P_n A : X \to X$ are uniformly bounded,*

$$\|A_n^{-1} P_n A\| \leq M. \tag{3.2}$$

The error estimate is valid

$$\|\varphi_n - \varphi\| \leq (1 + M) \inf_{\psi \in X_n} \|\psi - \varphi\|. \tag{3.3}$$

According to 11, (3.3) is usually called *the quasioptimal estimate* and corresponds to the error estimate of a projection method based on the approximation of elements from X by elements of subspaces X_n. It shows that the error of the projection method is determined by the quality of approximation of the exact solution by elements of subspace X_n.

Consider the equation

$$S\varphi + K\varphi = f \tag{3.4}$$

and the projection methods

$$P_n S\varphi_n = P_n f \tag{3.5}$$

and

$$P_n(S+K)\varphi_n = P_n f \tag{3.6}$$

with subspaces X_n and projection operators $P_n : Y \to Y_n$.

Lemma 12 *Assume that $S : X \to Y$ is a bounded operator with a bounded inverse $S^{-1} : Y \to X$ and that the projection method (3.5) is convergent for S. Let $K : X \to Y$ be a compact bounded operator and $S + K$ is injective. Then, the projection method (3.6) converges for $S + K$.*

Lemma 13 *Assume that $Y_n = S(X_n)$ and $\|P_n K - K\| \to 0$, $n \to \infty$. Then, for sufficiently large n, approximate equation (3.6) is uniquely solvable and the error estimate*

$$\|\varphi_n - \varphi\| \le M\|P_n S\varphi - S\varphi\|$$

is valid, where M is a positive constant depending on S and K.

For operator equations in Hilbert spaces, the projection method employing an orthogonal projection into finite-dimensional subspaces is called the Galerkin method.

Consider a pair of Hilbert spaces X and Y and assume that $A : X \to Y$ is an injective operator. Let $X_n \subset X$ and $Y_n \subset Y$ be the subspaces with $\dim X_n = \dim Y_n = n$. Then, $\varphi_n \in X_n$ is a solution of the equation $A\varphi = f$ obtained by the projection method generated by X_n and operators of orthogonal projections $P_n : Y \to Y_n$ if and only if

$$(A\varphi_n, g) = (f, g) \tag{3.7}$$

for all $g \in Y_n$. Indeed, equations (3.7) are equivalent to $P_n(A\varphi_n - f)$.

Equation (3.7) is called *the Galerkin equation.*

Assume that X_n and Y_n are the spaces of linear combinations of the basis and probe functions: $X_n = \{u_1, \ldots, u_n\}$ and $Y_n = \{v_1, \ldots, v_n\}$. Then, we can express φ_n as a linear combination

$$\varphi_n = \sum_{k=1}^{n} \gamma_n u_n,$$

and show that equation (3.7) is equivalent to the linear system of order n

$$\sum_{k=1}^{n} \gamma_k (Au_k, v_j) = (f, v_j), \quad j = 1, \ldots, n, \tag{3.8}$$

with respect to coefficients $\gamma_1, \ldots, \gamma_n$.

Thus, the Galerkin method may be specified by choosing subspaces X_n and projection operators P_n (or the basis and probe functions). We note that the convergence of the

Galerkin method do not depend on a particular choice of the basis functions in subspaces X_n and probe functions in subspaces Y_n. In order to perform a theoretical analysis of the Galerkin methods, we can choose only subspaces X_n and projection operators P_n. However, in practice, it is more convenient to specify first the basis and probe functions, because it is impossible to calculate the matrix coefficients using (3.8).

Now, let us apply the projection scheme of the Galerkin method to logarithmic integral equation

$$(L+K)\varphi = \int_{-1}^{1} (\ln \frac{1}{|x-y|} + K(x,y))\,\varphi(x)\frac{dx}{\sqrt{1-x^2}} = f(y), \quad -1 < y < 1. \tag{3.9}$$

Here, $K(x,y)$ is a smooth function; accurate conditions concerning the smoothness of $K(x,y)$ that guarantee the compactness of integral operator K are imposed in Section 1.2.

One can determine an approximate solution of equation (3.9) by the Galerkin method taking the Chebyshev polynomials of the first kind as the basis and probe functions. According to the scheme of the Galerkin method, we set

$$\varphi_N(x) = \frac{a_0}{2}T_0(x) + \sum_{n=1}^{N-1} a_n T_n(x),$$

$$X_N = \{T_0, \ldots, T_{N-1}\}, \quad Y_N = \{T_0, \ldots, T_{N-1}\}, \tag{3.10}$$

and find the unknown coefficients from finite linear system (3.8). Since $L : X_N \to Y_N$ is a bijective operator [see (1.51)], the convergence of the Galerkin method follows from Lemma 13 applied to operator L. If operator K is compact and $L+K$ is injective, then we can also prove the convergence of the Galerkin method for operator $L+K$ using Lemma 12. Note also that quasioptimal estimate 3.3 in the $L_2^{(1)}$ norm is valid for $\varphi_N(x) \to \varphi(x)$, $N \to \infty$.

Theorem 26 *If operator $L+K : L_2^{(1)} \to \widetilde{W}_2^1$ is injective and operator $K : L_2^{(1)} \to \widetilde{W}_2^1$ is compact, then the Galerkin method (3.10) converges for operator $L+K$.*

The same technique can be applied to constructing an approximate solution to the hypersingular equation

$$(H+K)\varphi = \frac{d}{dy}\int_{-1}^{1} \left(\frac{1}{x-y} + K(x,y)\right)\varphi(x)\sqrt{1-x^2}dx = f(y), \quad -1 < y < 1. \tag{3.11}$$

The conditions concerning the smoothness of $K(x,y)$ that guarantee the compactness of integral operator K are imposed in Section 1.2.

Choose the Chebyshev polynomials of the second kind as the basis and probe functions, i.e., set

$$X_N = \{U_0, \ldots, U_{N-1}\}, \quad Y_N = \{U_0, \ldots, U_{N-1}\}. \tag{3.12}$$

Unknown approximate solution $\varphi_N(x)$ is determined in the form

$$\varphi_N(x) = \sum_{n=0}^{N-1} b_n U_n(x). \tag{3.13}$$

Formula (1.52) shows that hypersingular operator H maps X_N onto Y_N and is bijective.

From Lemma 13, it follows that the Galerkin method for operator H converges. By virtue of Lemma 12, this result is also valid for operator $H+K$ under the assumption that K is compact and $H+K$ is injective. Quasioptimal formula (3.3) estimates the rate of convergence of approximate solution $\varphi_N(x)$ to exact one $\varphi(x)$, $N \to \infty$ in the norm of space $\widehat{W}_2^1$. Unknown coefficients a_n are found from finite linear system (3.8).

Theorem 27 *If operator $H+K : \widehat{W}_2^1 \to L_2^{(2)}$ is injective and $K : \widehat{W}_2^1 \to L_2^{(2)}$ is compact, then the Galerkin method (3.12) converges for operator $H+K$.*

3.3 Summation operators in the spaces of sequences

In this section, we will employ the approach based on the use of Chebyshev polynomials to construct families of summation operators associated with logarithmic integral equations.

Let us introduce the linear space h_p formed by the sequences of complex numbers $\{a_n\}$ such that

$$\sum_{n=1}^{\infty} |a_n|^2 n^p < \infty, \quad p \geq 0,$$

with the inner product and the norm

$$(a,b) = \sum_{n=1}^{\infty} a_n \bar{b}_n n^p, \quad \|a\|_p = \left(\sum_{n=1}^{\infty} |a_n|^2 n^p\right)^{\frac{1}{2}}.$$

One can easily verify the validity of the following statement.

Theorem 28 ([44], [50], [56]) *h_p, $p \geq 0$ is the Hilbert space.*

Let us introduce the summation operators

$$L = \begin{pmatrix} 1 & 0 & \dots & 0 & \dots \\ 0 & 2^{-1} & \dots & 0 & \dots \\ \dots & \dots & \dots & \dots & \dots \\ 0 & \dots & n^{-1} & 0 & \dots \\ \dots & \dots & \dots & \dots & \dots \end{pmatrix}$$

or, in other notation,

$$L = \{l_{nj}\}_{n,j=1}^{\infty}, \quad l_{nj} = 0, \; n \neq j, \quad l_{nn} = \frac{1}{n}, \; n = 1, 2, \dots,$$

and

$$L^{-1} = \begin{pmatrix} 1 & 0 & \dots & 0 & \dots \\ 0 & 2 & \dots & 0 & \dots \\ \dots & \dots & \dots & \dots & \dots \\ 0 & \dots & n & 0 & \dots \\ \dots & \dots & \dots & \dots & \dots \end{pmatrix}$$

or

$$L^{-1} = \{l_{nj}\}_{n,j=1}^{\infty}, \quad l_{nj} = 0, \; n \neq j, \quad l_{nn} = n, \; n = 1, \dots.$$

that play an important role in the theory of integral operators and operator-valued functions with a logarithmic singularity of the kernel.

Theorem 29 ([32], [44]) *L is a linear bounded invertible operator acting from the space h_p to the space h_{p+2}: $L : h_p \to h_{p+2}$ ($\Im L = h_{p+2}$, $\ker L = \{\emptyset\}$). The inverse operator $L^{-1} : h_{p+2} \to h_p$, is bounded, $p \geq 0$.*

□ Let us show that L acts from h_p to h_{p+2} and L^{-1} acts from h_{p+2} to h_p. Assume that $a \in h_p$. Then, for La we obtain

$$\sum_{n=1}^{\infty} \left|\frac{a_n}{n}\right|^2 n^{p+2} = \sum_{n=1}^{\infty} |a_n|^2 n^p < \infty.$$

If $b \in h_{p+2}$, we obtain similar relationships for $L^{-1}b$:

$$\sum_{n=1}^{\infty} |b_n n|^2 n^p = \sum_{n=1}^{\infty} |b_n|^2 n^{p+2} < \infty.$$

The norm $\|La\|_{p+2} = \|a\|_p$. The element $a = L^{-1}b$, $a \in h_p$, is the unique solution of the equation $La = b$, and $\ker L = \{\emptyset\}$. According to the Banach theorem, L^{-1} is a bounded operator. □

Note that $\|L\| = \|L^{-1}\| = 1$ and

$$L^{-1}L = LL^{-1} = E,$$

where E is the infinite unit matrix.

Assume that the summation operator A acting in h_p is given by its matrix $\{a_{nj}\}_{n,j=1}^{\infty}$ and the following condition holds:

$$\sum_{n=1}^{\infty}\sum_{j=1}^{\infty} \frac{n^{p+2}}{j^p}|a_{nj}|^2 \leq C^2 < \infty, \quad C = \text{const}, \ \ \mathrm{C} > 0.$$

Then, we can prove the following statement.

Theorem 30 ([32], [44]) *Operator $A : h_p \to h_{p+2}$ is completely continuous*

□ Let $\xi \in h_p$ and $A\xi = \eta$. Then,

$$\sum_{n=1}^{\infty} n^{p+2}|\eta_n|^2 = \sum_{n=1}^{\infty} n^{p+2}\left|\sum_{j=1}^{\infty} a_{nj}\xi_j\right|$$

$$\leq \|\xi\|_p^2 \sum_{n=1}^{\infty}\sum_{j=1}^{\infty} n^{p+2}\frac{|a_{nj}|^2}{j^p} \leq C^2\|\xi\|_p^2,$$

hence, $\eta \in h_{p+2}$. Now we will prove that A is completely continuous. In fact, operator A can be represented as a limit with respect to the operator norm of a sequence of finite-dimensional operators A_N, which are produced by the cut matrix A and defined by the matrices $A_N = \{a_{nj}\}_{n,j=1}^{N}$:

$$\|(A - A_N)\xi\|_{p+2}^2 = \sum_{n=1}^{N-1} n^{p+2}\left|\sum_{j=N}^{\infty} a_{nj}\xi_j\right|^2 + \sum_{n=N}^{\infty} n^{p+2}\left|\sum_{j=1}^{\infty} a_{nj}\xi_j\right|^2$$

$$\leq \|\xi\|_p^2 \sum_{n=N}^{\infty}\sum_{j=1}^{\infty} n^{p+2}\frac{|a_{nj}|^2}{j^p} + \|\xi\|_p^2 \sum_{n=1}^{N-1}\sum_{j=N}^{\infty} \frac{|a_{nj}|^2}{j^p}n^{p+2}$$

$$\leq \|\xi\|_p^2 \Big[\sum_{n=N}^{\infty}\Big(\sum_{j=1}^{\infty} n^{p+2}\frac{|a_{nj}|^2}{j^p}\Big) + \sum_{j=N}^{\infty}\Big(\sum_{n=1}^{\infty} n^{p+2}\frac{|a_{nj}|^2}{j^p}\Big)\Big],$$

and, consequently, the following estimates are valid with respect to the operator norm:

$$\|A - A_N\| \leq \sum_{n=N}^{\infty}\Big(\sum_{j=1}^{N} n^{p+2}\frac{|a_{nj}|^2}{j^p}\Big) + \sum_{j=N}^{\infty}\Big(\sum_{n=1}^{\infty} n^{p+2}\frac{|a_{nj}|^2}{j^p}\Big) \to 0, \quad N \to \infty. \quad \square$$

Corollary 3 *The sequence of operators A_N strongly converges to operator A: $\|A - A_N\| \to 0$, $N \to \infty$, and $\|A\| < C$, $C = \text{const}$. In order to provide the complete continuity of A, it is sufficient to require the fulfilment of the estimate*

$$\sum_{n=1}^{\infty}\sum_{j=1}^{\infty} n^{p+2}|a_{nj}|^2 < C, \quad C = \text{const}. \tag{3.14}$$

Definition 11 *Assume that condition (3.14) holds for the matrix elements of summation operator A. Then, the summation operator $F = L + A : h_p \to h_{p+2}$ is called the L-operator.*

Theorem 31 **([32], [43], [44])** *L-operator $F = L + A : h_p \to h_{p+2}$ is a Fredholm operator;* $\operatorname{ind}(L + A) = 0$.

Consider an L-operator-valued function

$$F(\lambda) = L + A(\lambda) : \ h_p \to h_{p+2}, \quad \lambda \in G,$$

where G is an open connected domain of the complex plane λ. We will assume that for $\lambda \in G$, $F(\lambda)$ is a holomorphic operator-valued function.

Definition 12 *The set $\rho(F) = \{\lambda$: there exists an inverse operator $F^{-1}(\lambda) : h_p \to h_{p+2}\}$ is called the resolvent set of operator-valued function $F(\lambda)$, and $\sigma(F) = G \setminus \rho(F)$ is called the spectrum of $F(\lambda)$.*

According to the definitions of the spectral theory of operator-valued functions [35] and [36] summarized in Section 1.6, the points of the spectrum of Fredholm operator-valued function $F(\lambda)$ are its CNs.

Assume that the matrix elements $\{a_{nj}(\lambda)\}_{n,j=1}^{\infty}$ of summation operator-valued function $F(\lambda)$ satisfy the following conditions:

(i) $a_{nj}(\lambda)$ are holomorphic in G;

(ii) $\sum_{n,j=1}^{\infty} j^{-p} n^{p+2}\|a_{nj}(\lambda)\|^2 \leq C^2(\lambda)$, and $C(\lambda)$ is continuous in G;

(iii) $\rho(F) \neq \emptyset$, i.e., in domain G, there exists at least one regular point of $F(\lambda)$.

Then, we can prove that the above conditions are sufficient to provide the fulfilment of the following statement.

Theorem 32 **([32], [44])** *Let the summation operator-valued function $F(\lambda)$ satisfy the conditions* (i)–(iii). *Then, $F(\lambda)$ is a holomorphic and Fredholm operator-valued function in G, and $\sigma(F)$ forms a discrete set without finite accumulation points in G.*

In order to prove that $F(\lambda)$ is (strongly) holomorphic operator-valued function in G (with respect to the norm) it is sufficient to verify [50] that $F(\lambda)$ is a bounded operator-valued function on any compact set in G and $F(\lambda)$ is a weakly holomorphic operator-valued function in G on fundamental sets. In other words, it is sufficient to verify that $(F(\lambda)u, g)$ is a holomorphic function in G for each element u from the fundamental set X in h_p and for each element g from the fundamental set Y in h_{p+2}. Recall that according to the definition [6], an arbitrary linear combinations of the elements of fundamental set X forms a set, which is dense in h_p, and an arbitrary linear combinations of the elements of fundamental set Y forms a set, which is dense in h_{p+2}.

First, it follows from condition *(ii)* that every point $\lambda \in G$ has a vicinity, where $F(\lambda)$ is bounded operator-valued function on any compact set in G. Second, choose fundamental set X as linear combinations of basic elements u_n from h_p and Y as linear combinations of basic elements g_n from h_{p+2}, where the nth coordinates of elements u_n and g_n are equal to 1 and other coordinates are equal to 0. Due to condition (i), $a_{nj}(\lambda) = (F(\lambda)u_n, g_j)$ is a holomorphic functions. Then, $(F(\lambda)u, g)$ is a holomorphic function for each element u from the fundamental set X in h_p and for each element g from the fundamental set Y in h_{p+2}. Hence, $F(\lambda)$ is a holomorphic operator-valued function.

Futhermore, condition (ii) implies the fulfilment of condition (3.14). It follows from Corollary 3 that $A(\lambda)$ is a compact operator. Hence, $F(\lambda) = L + A(\lambda)$ is a Fredholm operator as a sum of continuously invertible and complete continuous operators.

Then, condition (iii) yields [50] that $\sigma(F)$ forms a discrete set without finite accumulation points in G.

3.4 Matrix representation of logarithmic integral operators and operator-valued functions

Consider a class Φ of functions φ such that they admit in the interval $(-1, 1)$ the representation in the form of the Fourier–Chebyshev series

$$\varphi(x) = \frac{1}{2}T_0(x)\xi_0 + \sum_{n=1}^{\infty} \xi_n T_n(x); \quad -1 \le x \le 1, \tag{3.15}$$

where $T_n(x) = \cos(n \arccos x)$ are the Chebyshev polynomials and $\xi = (\xi_0, \xi_1, ...\xi_n, ...) \in h_2$; that is,

$$\|\xi\|_2^2 = \frac{1}{2}|\xi_0|^2 + \sum_{n=1}^{\infty} |\xi_n|^2 n^2 < \infty.$$

In this section we add the coordinate ξ_0 to vector ξ and modify the corresponding formulas. It is easy to see that all results of previous sections are also valid in this case. Recall the definitions of weighted Hilbert spaces associated with logarithmic integral operators (see Section 1.3):

$$L_2^{(1)} = \{f(x) : \|f\|_1^2 = \int_{-1}^{1} |f(x)|^2 \frac{dx}{\sqrt{1-x^2}} < \infty\},$$

$$L_2^{(2)} = \{f(x) : \|f\|_2^2 = \int_{-1}^{1} |f(x)|^2 \sqrt{1-x^2}\, dx < \infty\},$$

$$\widetilde{W}_2^1 = \{f(x) : \ f \in L_2^{(1)}, \ f' \in L_2^{(2)}\},$$

and

$$\widehat{W}_2^1 = \{f(x) : \ f \in L_2^{(2)}, f^*(x) = f(x)\sqrt{1-x^2}, \ f^{*\prime} \in L_2^{(2)}, \ f^*(-1) = f^*(1) = 0\}.$$

Below, we will denote the weights by $p_1(x) = \sqrt{1-x^2}$ and $p_1^{-1}(x)$.

Note that the Chebyshev polynomials specify orthogonal bases in spaces $L_2^{(1)}$ and $L_2^{(2)}$. In order to obtain the matrix representation of the integral operators with a logarithmic singularity of the kernel, we will prove a similar statement for weighted space $\widetilde{W}_2^1$.

Lemma 14 *$\widetilde{W}_2^1$ and Φ are isomorphic.*

□ Assume that $\varphi \in \Phi$. We will show that $\varphi \in \widetilde{W}_2^1$. Function $\varphi(x)$ is continuous, and (3.15) is its Fourier–Chebyshev series with the coefficients

$$\xi_n = \frac{2}{\pi} \int_{-1}^{1} T_n(x)\varphi(x)p_1^{-1}(x)dx.$$

The series $\sum_{n=1}^{\infty} |\xi_n|^2 n^2$ converges and, due to the Riesz–Fischer theorem [50], there exists an element $f \in L_2^{(1)}$, such that its Fourier coefficients in the Chebyshev polynomial series are equal to $n\xi_n$; that is, setting $U_n(x) = \sin(n \arccos x)$, one obtains

$$n\xi_n = \frac{2}{\pi} \int_{-1}^{1} nT_n(x)\varphi(x)p_1^{-1}(x)dx = -\frac{2}{\pi} \int_{-1}^{1} U_n'(x)\varphi(x)dx$$

$$= -\frac{2}{\pi} \int_{-1}^{1} U_n'(x)(p_1(x)\varphi(x))\frac{dx}{p_1(x)} = \frac{2}{\pi} \int_{-1}^{1} U_n(x)f(x)\frac{dx}{p_1(x)}, \quad n \geq 1.$$

Since $f(x)$ is an integrable function, the integral $\int_{-1}^{1} |f(x)|p_1^{-1}(x)dx$ converges, and

$$F(x) = \int_{-1}^{x} \frac{f(t)dt}{p_1(t)}$$

is absolutely continuous (as a function in the form of an integral depending on the upper limit of integration), $F'(x) = f(x)p_1^{-1}(x)$, and

$$\frac{2}{\pi} \int_{-1}^{1} U_n(x)f(x)\frac{dx}{p_1(x)} = \frac{2}{\pi} \int_{-1}^{1} U_n(x)F'(x)dx = -\frac{2}{\pi} \int_{-1}^{1} U_n'(x)F(x)dx.$$

The latter equalities yield

$$\int_{-1}^{1} U_n'(x)[\varphi(x) - F(x)]dx = -n \int_{-1}^{1} T_n(x)[\varphi(x) - F(x)]\frac{dx}{p_1(x)} = 0, \ n \geq 1.$$

Hence, $\varphi(x) - F(x) = C =$ const. The explicit form of $F'(x)$ yields $\varphi'(x) = f(x)p_1^{-1}(x)$; therefore, $\varphi'(x)p_1(x) \in L_2^{(1)}$ and $\varphi'(x) \in L_2^{(2)}$.

Now, let us prove the inverse statement. Let $\varphi \in \widetilde{W}_2^1$ and be represented by the Fourier–Chebyshev series (3.15). Form the Fourier series

$$p_1(x)\varphi'(x) = \sum_{n=1}^{\infty} \eta_n U_n(x), \quad |x| \le 1,$$

where

$$\eta_n = \frac{2}{\pi}\int_{-1}^{1} U_n(x)(p_1(x)\varphi'(x))\frac{dx}{p_1(x)} = -\frac{2}{\pi}\int_{-1}^{1} U_n'(x)\varphi(x)dx = \frac{2n}{\pi}\int_{-1}^{1} T_n(x)\frac{\varphi(x)dx}{p_1(x)}.$$

Now, in order to prove that $\xi \in h_2$, it is sufficient to apply the Bessel inequality to $\eta_n = n\xi_n$, where

$$\xi_n = \frac{2}{\pi}\int_{-1}^{1} \varphi(x)T_n(x)\frac{dx}{p_1(x)}.$$

□

Consider the integral operator with a logarithmic singularity of the kernel

$$L\varphi = -\frac{1}{\pi}\int_{-1}^{1} \ln|x-s|\varphi(s)\frac{ds}{p_1(s)}, \quad \varphi \in \widetilde{W}_2^1,$$

and the integral equation

$$L\varphi + N\varphi = f, \quad \varphi \in \widetilde{W}_2^1, \tag{3.16}$$

where

$$\Phi_N(x) = N\varphi = \int_{-1}^{1} N(\mathbf{q})\varphi(s)\frac{ds}{\pi p_1(x)}, \quad \mathbf{q} = (x, s),$$

and $N(\mathbf{q})$ is an arbitrary complex-valued function satisfying the continuity conditions

$$N(\mathbf{q}) \in C^1(\Pi_1); \quad \frac{\partial^2 N}{\partial x^2}(\mathbf{q}) \in L_2[\Pi_1, p_1^{-1}(x)p_1^{-1}(s)],$$

$$\Pi_1 = ([-1,1] \times [-1,1]).$$

Write the Fourier–Chebyshev expansions

$$\varphi(s) = \frac{1}{\sqrt{2}}\xi_0 T_0 + \sum_{n=1}^{\infty} \xi_n T_n(s), \quad \xi = (\xi_0, \xi_1, \ldots, \xi_n, \ldots) \in h_2, \tag{3.17}$$

and

$$f(x) = \frac{f_0}{2}T_0(x) + \sum_{n=1}^{\infty} f_n T_n(x),$$

where the coefficients

$$f_n = \frac{2}{\pi}\int_{-1}^{1} f(x)T_n(x)\frac{dx}{\sqrt{1-x^2}}, \quad n = 0, 1, \ldots.$$

Function $\Phi_N(x)$ is continuous and may be thus also expanded in the Fourier–Chebyshev series

$$\Phi_N(x) = \frac{1}{2}b_0T_0(x) + \sum_{n=1}^{\infty} b_nT_n(x), \tag{3.18}$$

where

$$b_n = 2\int\int_{\Pi_1} N(\mathbf{q})T_n(x)\varphi(s)\frac{dxds}{\pi^2 p_1(x)p_1(s)}. \tag{3.19}$$

According to Lemma 14 and formulas (3.17) and (3.18), equation (3.16) is equivalent in $\widetilde{W}_2^1$ to the equation

$$\begin{aligned} \ln 2\frac{\xi_0}{\sqrt{2}}T_0(x) + \sum_{n=1}^{\infty} n^{-1}\xi_nT_n(x) + \frac{b_0}{2}T_0(x) + \sum_{n=1}^{\infty} b_nT_n(x) \\ = \frac{f_0}{2}T_0(x) + \sum_{n=1}^{\infty} f_nT_n(x), \quad |x| \le 1. \end{aligned} \tag{3.20}$$

Note that since $\Phi_N(x)$ is a differentiable function of x in $[-1,1]$, the series in (3.20) converge uniformly, and the equality in (3.20) is an identity in $(-1,1)$.

Now, we substitute series (3.17) for $\varphi(x)$ into (3.19). Taking into account that $\{T_n(x)\}$ is a basis and one can change the integration and summation and to pass to double integrals (these statements are verified directly), we obtain an infinite system of linear algebraic equations

$$\begin{cases} \sqrt{2}\ln 2\xi_0 + b_0 = f_0, \\ n^{-1}\xi_n + b_n = f_n, \qquad n \ge 1. \end{cases} \tag{3.21}$$

In terms of the summation operators, this systems can be represented as

$$L\xi + A\xi = f; \quad \xi = (\xi_0, \xi_1, \dots, \xi_n, \dots) \in h_2; \quad f = (f_0, f_1, \dots, f_n, \dots) \in h_4.$$

Here, operators L and A are defined by

$$L = \{l_{nj}\}_{n,j=0}^{\infty}, \quad l_{nj} = 0,\ n \ne j; \quad l_{00} = \ln 2; \quad l_{nn} = \frac{1}{n},\ n = 1,2,\dots;$$

$$A = \{a_{nj}\}_{n,j=0}^{\infty}, \quad a_{nj} = \varepsilon_{nj}\int\int_{\Pi_1} N(\mathbf{q})T_n(x)T_j(s)\frac{dxds}{\pi^2 p_1(x)p_1(s)}, \tag{3.22}$$

where

$$\varepsilon_{nj} = \begin{cases} 1, & n = j = 0, \\ 2, & n \le 1,\ j \le 1, \\ \sqrt{2}, & \text{for other } n, j. \end{cases}$$

Lemma 15 *Let $N(\mathbf{q})$ satisfy the continuity conditions formulated in (3.16). Then*

$$\sum_{n=1,j=0}^{\infty} |a_{nj}|^2 n^4 + \sum_{j=0}^{\infty} |a_{0j}|^2 \frac{1}{\ln^2 2} \le C_0, \tag{3.23}$$

where

$$C_0 = \frac{1}{\ln^2 2}\int_{-1}^{1}\left|\int_{-1}^{1} \frac{N(\mathbf{q})dx}{\pi p_1(x)}\right|^2 \frac{ds}{\pi p_1(s)} + \int\int_{\Pi_1} p_1^2(x)\,|(p_1(x)N_x')_x'|^2\,\frac{dxds}{\pi^2 p_1(x)p_1(s)}. \tag{3.24}$$

□ Let us expand a function of two variables belonging to the weighted space $L_2[\Pi_1, p_1^{-1}(x)p_1^{-1}(s)]$ in the double Fourier–Chebyshev series applying the Fourier–Chebyshev system $\{T_n(x)T_m(s)\}$:

$$-p_1(x)(p_1(x)N'_x(\mathbf{q}))'_x = \sum_{n=0}^{\infty}\sum_{m=0}^{\infty} C_{nm}T_n(x)T_m(s),$$

where

$$\begin{aligned} C_{nm} &= -\frac{\varepsilon_{nm}^2}{\pi}\int_{-1}^{1} T_m(s)\frac{ds}{p_1(s)}\int_{-1}^{1} T_n(x)(p_1(x)N'_x(\mathbf{q}))'_x dx \\ &= \frac{\varepsilon_{nm}^2}{\pi}\int_{-1}^{1} T_m(s)\frac{ds}{p_1(s)}\int_{-1}^{1} T'_n(x)p_1(x)N'_x(\mathbf{q})dx \\ &= \frac{n\varepsilon_{nm}^2}{\pi^2}\int_{-1}^{1}\frac{T_m(s)}{p_1(s)}ds\int_{-1}^{1} U_n(x)N'_x(\mathbf{q})dx \\ &= \frac{n^2\varepsilon_{nm}^2}{\pi^2}\iint_{\Pi_1}\frac{N(\mathbf{q})T_n(x)T_m(s)}{p_1(x)p_1(s)}dxds = n^2\varepsilon_{nm}a_{nm}. \end{aligned}$$

The Bessel inequality yields

$$\sum_{n=1}^{\infty}\sum_{m=0}^{\infty} n^4\pi^2|a_{nm}|^2 \le \iint_{\Pi_1} p_1^2(x)|[p_1(x)N'_x(\mathbf{q})]'_x|^2\frac{dxds}{p_1(x)p_1(s)}.$$

Let us expand the function

$$g(s) = \frac{1}{\ln 2}\int_{-1}^{1}\frac{N(\mathbf{q})dx}{\pi^2 p_1(x)}$$

in the Fourier–Chebyshev series. Again, due to the Bessel inequality, we obtain

$$\frac{1}{\ln^2 2}\sum_{m=0}^{\infty}|a_{0m}|^2 \le \frac{1}{\ln^2 2}\int_{-1}^{1}\left|\int_{-1}^{1}\frac{N(\mathbf{q})ds}{\pi p_1(x)}\right|^2\frac{ds}{\pi p_1(s)}. \quad \square$$

Assume now that the continuous component $N = N(\lambda, \mathbf{q})$ of the integral operator in (3.16) is holomorphic with respect to complex variable λ in a domain G at every point $\mathbf{q} \in \Pi_1$. Then, the left-hand side of (3.16) defines an operator-valued function acting on functions from the Hilbert space $H = \widetilde{W}_2^1$. Consider the problem on characteristic numbers for the integral operator-valued function $K(\lambda) = L + N(\lambda)$ with a logarithmic singularity of the kernel:

$$L\varphi + N(\lambda)\varphi = 0, \quad \varphi \in \widetilde{W}_2^1, \tag{3.25}$$

where

$$N(\lambda)\varphi = \int_{-1}^{1} N(\lambda, \mathbf{q})\varphi(s)\frac{ds}{\pi p_1(s)}$$

Applying a similar procedure (described in detail in [31] , [56] and [44]), one can reduce (3.25) to an equivalent summation equation

$$K_L(\lambda)\xi \equiv L\xi + A(\lambda)\xi = 0, \quad \xi \in h_2, \quad A = \|a_{nm}(\lambda)\|_{n,m=0}^{\infty},$$

which is the problem on CNs for an L-operator-valued function K_L. Here, infinite matrix L and the matrix element of A are determined by the correponding formulas (3.22). According to [44], in this case, $a_{mn} = a_{mn}(\lambda)$ are holomorphic in G and estimate (3.23) remains valid with $C_0 = C_0(\lambda)$ in (3.24) being a (bounded positive) continuous function of λ in G. The latter properties of A and the application of Corollary 2 yield the following statement.

Theorem 33 ([32], [44], [56]) *$K(\lambda)$ and $K_L(\lambda)$ are Fredholm operator-valued functions holomorphic in G, and they may have not more than a finite number of CNs in every closed bounded domain $G_0 \subset G$.*

A detailed procedure of the determination of CNs of the summation operator-valued functions $K_L(\lambda)$ is described in [26, 27].

Chapter 4

Boundary value problems reduced to logarithmic integral equations

In this chapter, we summarize the main results concerning the solvability of two-dimensional problems for the Helmholtz equation. We will limit ourselves to a brief survey, assuming that details can be found in original monographs. We will consider the boundary value problems that arise in electromagnetics and acoustics when the plane wave diffraction and free oscillations (eigenoscillations) of the electromagnetic or acoustic fields in open and closed cylindrical domains are studied.

First, we focus on mathematical statements of the problem of diffraction by obstacles with curved boundaries; the solutions are represented in the form of single- and double-layer potentials, and pseudodifferential equations on the curve are derived. The solvability theory is constructed for pseudodifferential equations and boundary value problems in the Sobolev spaces. Then, we consider the problems of diffraction by closed curves. For unclosed curves, it is natural to begin the analysis of the diffraction problems with an important particular case when the boundary of the obstacle is formed by rectilinear intervals (strips). Note that for such a rectilinear boundary, the spaces of solutions and images and PDOs can be described explicitly in terms of the Fourier transforms.

4.1 Formulation of the problems

Denote by Γ the boundaries of the cross sections [in the plane $\mathbf{x} = (x, y)$] of cylindrical domains with the generatrix directed along the z axis and consider the following three typical cases.

(i) Γ is an unclosed nonintersecting finite piecewise smooth curve (a strip) situated in the free space (in particular, strip Γ may coincide with a rectilinear segment or a circular arc, and such open resonators and scatteres are studied, e.g., in [15, 22, 24, 46]). We will also consider a more general case when Γ consists of a finite number of simple, planar, closed or unclosed smooth curves Γ_j of a finite length that belong to class C^∞. Introduce the following notation:

$$\Gamma = \bigcup_j \Gamma_j, \quad \overline{\Gamma}_i \bigcap \overline{\Gamma}_j = \emptyset \ \text{ for } \ i \neq j,$$

where the edges

$$\partial\Gamma = \bigcup_j (\overline{\Gamma}_j \setminus \Gamma_j)$$

are the endpoints of Γ (these points do not belong to Γ and $\partial\Gamma \bigcap \Gamma = \emptyset$). Let G_0 be a union of internal domains bounded by closed curves Γ_j, and we set $G_0 = \emptyset$ if these domains are absent.

(ii) $\Gamma = (\partial\Omega^1 \cup \partial\Omega^2) \setminus \overline{\Sigma}$, where $\partial\Omega^1$ and $\partial\Omega^2$ are the boundaries of the half-plane $\Omega^1 = \Omega^+ = \{\mathbf{x} : y > 0\}$ [which is, of course, an unbounded $S_\Pi(\Sigma)$-domain introduced in Chapter 2] and of a bounded $S_\Pi(\Sigma)$-domain $\Omega^2 \subset \Omega^- = \{\mathbf{x} : y < 0\}$, respectively, having a common part $\Gamma_{12} = (\partial\Omega^2 \cap \partial\Omega^1) \in \{\mathbf{x} : y = 0\}$ containing an interval (a slot) Σ, $\Sigma \subseteq \Gamma_{12}$; function $\varepsilon = \varepsilon(\mathbf{x}) = \varepsilon_n = \text{const}\ \mathbf{x} \in \Omega^n, n = 1, 2$ specifies the permittivity of the medium. We indicate a particular case when the $S_\Pi(\Sigma)$-domain $\Omega^2 = \Pi_{ab} = \{\mathbf{x} : -a/2 < x < a/2 \ \ -b < y < 0\}$ is a rectangle, and $\Sigma = \{\mathbf{x} : y = 0,\ -l < x < l\}$ is the slot with the edges $\partial\Sigma = \{A_1 = -l,\ A_2 = l\}$.

(iii) $\Gamma = (\partial\Pi_{ab_1} \cup \partial\Pi_{ab_2}) \setminus \overline{L}$, where $\Omega^1 = \Pi_{ab_1} = \{\mathbf{x} =: 0 < x < a,\ 0 < y < b_1\}$ and $\Omega^2 = \Pi_{ab_2} = \{\mathbf{x} : 0 < x < a, b_2 < y < 0\}$ are two rectangular domains, and $\Sigma = \{\mathbf{x} : y = 0, a/2 - w = d_1 < x < d_2 = a/2 + w\}$ is the slot with the edges $\partial\Sigma = \{A_1 = d_1,\ A_2 = d_2\}$.

In case (i), the acoustic and electromagnetic two-dimensional problems of diffraction by such curved boundaries are reduced [61, 62] to the determination of a scalar function u (the scattered field) that satisfies the homogeneous Helmholtz equation

$$\Delta u + k^2 u = 0 \quad \text{in} \quad \mathrm{R}^2 \setminus (\overline{\Gamma} \bigcup G_0), \tag{4.1}$$

the Dirichlet,

$$u|_\Gamma = -f, \tag{4.2}$$

or the Neumann,

$$\left.\frac{\partial u}{\partial n}\right|_\Gamma = -g, \tag{4.3}$$

boundary conditions on Γ, and the Sommerfeld radiation conditions at infinity

$$u = O(r^{-1/2}), \quad \frac{\partial u}{\partial r} - iku = o(r^{-1/2}) \tag{4.4}$$

for $r = |\mathbf{x}| = (x^2 + y^2)^{1/2} \to \infty$. The Dirichlet problem corresponds to a soft screen in acoustics (or to the case of E-polarization in electrodynamics), and the Neumann problem, to a hard screen (or to the case of H-polarization). Here, k means the free-space wavenumber and $\Im k \geq 0$, $k \neq 0$.

In addition to the above conditions, field u must satisfy the requirement that provides the finiteness of energy in every bounded spatial domain:

$$u \in H^1_{loc}(\mathrm{R}^2 \setminus \overline{G_0}); \tag{4.5}$$

that is,

$$\int\limits_{G \setminus G_0} \left(|\nabla u|^2 + |u|^2\right) d\mathbf{x} < \infty$$

for every bounded domain G.

We also formulate the boundary value problems for the homogeneous Helmholtz equation with a piecewise constant coefficient

$$\Delta u(\mathbf{x}) + \lambda\varepsilon(\mathbf{x})u(\mathbf{x}) = 0, \quad \lambda = k^2, \quad \mathbf{x} \in \Omega, \tag{4.6}$$

where $\Omega = R^2 \setminus S$ in case (i) and $\Omega = \Omega^1 \cup \Omega^2$ in cases (ii) and (iii). For the classical solution, one may require additionally that

$$u \in M = \{u: \ u \in C^2(\Omega) \cap C^1(\cup_i \overline{\Omega}^{\varepsilon_i} \setminus \Gamma_\delta)\},$$

where

$$\Gamma_\delta = \{\mathbf{x}: \ dist(\mathbf{x}, \partial\Gamma) < \delta\}$$

is a δ-vicinity of the endpoints of curve Γ (edges), and $\Omega^{\varepsilon_j} = \{\mathbf{x}: \varepsilon(\mathbf{x}) = \varepsilon_j = \text{const}\}$; solution u satisfies homogeneous boundary conditions (4.2) or (4.3) on Γ, the conjugation conditions on Σ [in cases (ii) and (iii)]

$$u^+ - u^- = 0, \quad \left[\frac{1}{\varepsilon(\mathbf{x})}\frac{\partial u^+}{\partial n} - \frac{1}{\varepsilon_0}\frac{\partial u^-}{\partial n}\right] = 0, \tag{4.7}$$

or

$$u^+ - u^- = 0, \quad \left[\frac{\partial u^+}{\partial n} - \frac{\partial u^-}{\partial n}\right] = 0; \tag{4.8}$$

in case (ii), when the diffraction of the plane H- or E-polarized wave is considered, the first (homogeneous) conjugation condition may be replaced by the inhomogeneous condition

$$u^+ - u^- = \psi_1, \quad \left[\frac{1}{\varepsilon(\mathbf{x})}\frac{\partial u^+}{\partial n} - \frac{1}{\varepsilon_0}\frac{\partial u^-}{\partial n}\right] = 0, \tag{4.9}$$

or

$$u^+ - u^- = 0, \quad \left[\frac{\partial u^+}{\partial n} - \frac{\partial u^-}{\partial n}\right] = \psi_2, \tag{4.10}$$

where $\psi_j = \psi_j(x)$ $(j = 1, 2)$ is the given differentiable function on Σ and the Meixner (edge) conditions (4.5). For complex λ, the Sommerfeld radiation condition can be replaced by a more general radiation conditions [47]: there exists an $R_0 > 0$ such that for all $\mathbf{x} = (r, \phi): r \geq R_0$, the following representations hold

$$u(\mathbf{x}) = \sum_{m=-\infty}^{+\infty} a_m H_m^{(1)}(kr) p_m(\phi), \tag{4.11}$$

where a_m are the complex numbers, $p_m(\phi) = e^{im\phi}$ $(m = 0, \pm 1, \pm 2, \ldots)$ in case (i); in case (ii), $p_m(\phi) = 0$ $(m = -1, -2, \ldots)$ and $p_m(\phi) = \sin m\phi$ for (4.2) and $p_m(\phi) = \cos m\phi$ for (4.3); the series in (4.11) admit double termwise differentiation with respect to $r \geq R_0$ and $\phi \in (0, 2\pi)$ [case (i)] or $\phi \in (0, \pi)$ [case (ii)]; the Hankel function $H_m^{(1)}(z)$ of the first kind and mth order may be generally considered on the appropriate Riemann surface of complex variable z. of the analytical continuation with respect to the spectral parameters k or λ of the fundamental solution $\mathcal{E}$ of the two-dimensional Helmholtz operator $\mathcal{H}$; (in electromagnetic diffraction problems, $k = \frac{\omega}{c}$ is referred to as the wavenumber of free space, ω is the frequency parameter, and c is the speed of light in free space). Note that $\mathcal{E}(\mathbf{x} - \mathbf{y})$ satisfies the Sommerfeld radiaton conditions in domain Ω^+ for real values of k.

Homogeneous problems (4.1), (4.2) with $f = 0$, (4.5), and (4.11); (4.1), (4.3) with $g = 0$, (4.5), and (4.11); (4.6), (4.2) with $f = 0$, (4.8), (4.5), and (4.11); and (4.6), (4.3) with $g = 0$, (4.7), (4.5), and (4.11) concerning the determination of (complex) values of spectral parameters k or λ (eigenvalues), for which there exist their nontrivial solutions (in appropriate spaces), will be called *problems* E (with the Dirichlet conditions on Γ) and *problems* H (with the Neumenn conditions on Γ).

According to the terminology [24, 26, 31], these problems may be conditionally referred to as the problem on eigenfrequencies of open [(ii)] and shielded [(iii)] slot resonators.

The respective inhomogeneous boundary value problem (4.1), (4.2), (4.4), and (4.5); (4.1), (4.3)–(4.5); (4.6), (4.2), (4.8), (4.4), and (4.5) [or (4.6), (4.2) with $f = 0$, (4.10), (4.4), and (4.5)]; and (4.6), (4.3), (4.7), (4.4), and (4.5) [or (4.6), (4.3) with $g = 0$, (4.9), (4.4), and (4.5)] (that correspond in electromagnetics to the diffraction of the plane H-polarized wave), will be called, respectively, *problems Ei* and *problems Hi.*

The form of the radiation condition (4.11) in these problems is connected with the fact that the spectral parameter may be complex. For real k (when considering the diffraction problem Hi in the lossless medium with $\Im\varepsilon = 0$), one can replace (4.11) by the equivalent Sommerfeld radiation conditions (4.4) or the Kupradze radiation conditions [48] for the complex k with $\Im k > 0$. (4.11) may be considered as a generalization of the latter conditions and can be applied for arbitrary nonzero complex k (or λ). This condition may be also considered as the continuation of the Sommerfeld condition from the real axis of the complex parameter or of the Kupradze condition from the zero-sheet upper half-plane to the appropriate Riemannian surface.

In (4.2) and (4.3), the equality means the equality of elements from spaces $H^{1/2}(\Gamma)$ and $H^{-1/2}(\Gamma)$, and (4.3) must hold on both sides of Γ. It is well known [63] that solutions to the homogeneous Helmholtz equation belonging to $H^1_{loc}(\mathrm{R}^2 \setminus (\overline{\Gamma} \bigcup G_0))$ are infinitely differentiable in $\mathrm{R}^2 \setminus (\overline{\Gamma} \bigcup G_0)$; therefore, one can assume that $u \in C^2(\mathrm{R}^2 \setminus (\overline{\Gamma} \bigcup G_0))$ and understand (4.1) in a usual sense. If the field sources are situated outside the curved boundaries, then functions f and g, which correspond to the trace of the incident field and its normal derivative on Γ, are infinitely differentiable on Γ and $f, g \in C^\infty(\overline{\Gamma})$. Moreover, if $\Gamma' \subset \Gamma$ is a smooth part of a C^∞ curve, then one can use the regularity theorem [63] to show that solution u is infinitely differentiable up to Γ' (on each side of the curve if $L' \bigcap \partial G_0 = \emptyset$), and

$$u \in C^2(\mathrm{R}^2 \setminus (\overline{\Gamma} \bigcup G_0)) \bigcap_{\delta>0} C^1(\overline{G_1 \setminus (\Gamma_\delta \bigcup G_0)}) \bigcap_{\delta>0} C^1(\overline{G_2 \setminus \Gamma_\delta}),$$

if all Γ_j belong to C^∞. Note that, in [68], diffraction problem (4.1)–(4.5) is investigated, in the generalized formulation for a single smooth curve.

4.2 Uniqueness and existence theorems

Consider the uniqueness of the problems (4.1, (4.2), (4.4), and (4.5) and (4.1, (4.3)–(4.5). We will prove that for $\Im k \geq 0$ and $k \neq 0$, these Dirichlet and Neumann problems with homogeneous boundary conditions have only trivial solutions. Denote by $[\ \cdot\ ]$ the difference of the limiting values of a function taken from different sides of a curve. Equations (4.1)–(4.3) yield the transmission problem for u:

$$\Delta u + k^2 u = 0 \quad \text{in} \quad G_2 \bigcup (G_1 \setminus G_0),$$

$$[u]_{\Lambda \setminus \overline{\Gamma}} = \left[\frac{\partial u}{\partial n}\right]_{\Lambda \setminus \overline{\Gamma}} = 0,$$

$$u|_\Gamma = 0 \quad \text{or} \quad \left.\frac{\partial u}{\partial n}\right|_\Gamma = 0,$$

with the radiation condition (4.4) for $r \to \infty$. Recall that Λ is a closed, smooth C^∞ curve that contains Γ ($\Gamma \subset \Lambda$) and divides the plane into bounded, G_1, and unbounded, G_2, domains, and n is the normal vector external with respect to G_1.

Denote by $B_R = \{\mathbf{x} : |\mathbf{x}| < R\}$ a circle of radius R such that $\Lambda \subset B_R$. Applying the second Green's formula in domains $G_1 \setminus G_0$, $G_2 \bigcap B_R$ to functions u and $\overline{u}$ and adding up the results, we obtain

$$\int\limits_{\partial B_R} \left(\frac{\partial u}{\partial r}\overline{u} - \frac{\partial \overline{u}}{\partial r} u \right) dl = -4ik'k'' \int\limits_{B_R} |u|^2 \, d\mathbf{x}, \quad k = k' + ik''.$$

Integrals over Λ vanish by virtue of the boundary and transmission conditions. Here, the applicability of the Green's formula follows from the fact that u belongs to $H^1_{loc}(\mathrm{R}^2 \setminus G_0)$ and contour Λ is smooth [11]. A standard analysis of the obtained identity (see [61]) shows that $u \equiv 0$ in G_2. The transmission conditions and the fact that u is an analytical function in $\mathrm{R}^2 \setminus (\overline{\Gamma} \bigcup G_0)$ yield $u \equiv 0$ in $G_1 \setminus G_0$. Thus, the uniqueness of the solution to problem (4.1)–(4.5) is proved.

We will look for the unique solution to problem (4.1)–(4.5) in the form of potentials

$$u(\mathbf{x}) = K_0(q\varphi) = -\frac{i}{4} \int\limits_{\Gamma} H_0^{(1)}(k|\mathbf{x} - \mathbf{y}|)\varphi(\mathbf{y})\, dl, \quad \varphi \in \tilde{H}^{-1/2}(\Gamma) \tag{4.12}$$

for the Dirichlet problem and

$$u(\mathbf{x}) = K_1(q\psi) = \frac{i}{4} \int\limits_{\Gamma} \frac{\partial}{\partial n_y} H_0^{(1)}(k|\mathbf{x} - \mathbf{y}|)\psi(\mathbf{y})\, dl, \quad \psi \in \tilde{H}^{1/2}(\Gamma) \tag{4.13}$$

for the Neumann problem. Here, $H_0^{(1)}(z)$ is the Hankel function of the first kind. For the case of a closed contour, potentials (4.12) and (4.13) were considered in [11]. Continuing φ and ψ as zero functions from Γ to Λ, performing the restriction on Γ, and taking into account the continuity of operators q and p, we prove, using the results [11], that

$$\varphi = \left[\frac{\partial u}{\partial n}\right]_\Gamma \text{ and } [u]_\Gamma = 0 \text{ for the Dirichlet problem}$$

and

$$\psi = [u]_\Gamma \text{ and } \left[\frac{\partial u}{\partial n}\right]_\Gamma = 0 \text{ for the Neumann problem.}$$

In addition to this,

$$u \in H^1_{loc}(\mathrm{R}^2 \setminus \overline{G_0})$$

for all

$$\varphi \in \tilde{H}^{-1/2}(\Gamma), \quad \psi \in \tilde{H}^{1/2}(\Gamma)$$

[see (4.12) and (4.13)]. Here,

$$[u]_\Gamma = p\left(\gamma_0\left(u|_{G_2}\right) - \gamma_0\left(u|_{G_1}\right)\right),$$

$$\left[\frac{\partial u}{\partial n}\right]_\Gamma = p\left(\gamma_1\left(u|_{G_2}\right) - \gamma_1\left(u|_{G_1}\right)\right).$$

Since solution u is chosen in the form of potentials (4.12) and (4.13), equation (4.1) and conditions (4.4) and (4.5) hold for all φ and ψ. Thus, we have to satisfy only boundary conditions (4.2) or (4.3), which yields the equations on curves Γ:

$$D\varphi = \frac{i}{4}\int\limits_{\Gamma} H_0^{(1)}(k|\mathbf{x}-\mathbf{y}|)\varphi(\mathbf{y})\,dl = f(\mathbf{x}), \quad \mathbf{x}\in\Gamma, \tag{4.14}$$

$$D = p\gamma_0 K_0 q,$$

for the Dirichlet problem, and

$$N\psi = \frac{i}{4}\frac{\partial}{\partial n_{\mathbf{x}}}\int\limits_{\Gamma}\frac{\partial}{\partial n_{\mathbf{y}}} H_0^{(1)}(k|\mathbf{x}-\mathbf{y}|)\psi(\mathbf{y})\,dl = -g(\mathbf{x}), \quad \mathbf{x}\in\Gamma, \tag{4.15}$$

$$N = p\gamma_1 K_1 q,$$

for the Neumann problem. The first equation has a weak (logarithmic) singularity and the second equation is hypersingular. Using the methods of the analysis of logarithmic integral operators in the Sobolev spaces in terms of PDOs (see Section 1.8), we can consider both equations (4.14) and (4.15) from a single viewpoint as pseudodifferential equations.

Thus, if φ and ψ are solutions to (4.14) and (4.15), formulas (4.12) and (4.13) give the solution to diffraction problem (4.1)–(4.5).

Now, we consider the case when all curves Γ_j are closed and smooth. Then, according to [11],

$$D : H^{-1/2}(\Gamma) \to H^{1/2}(\Gamma)$$

and

$$N : H^{1/2}(\Gamma) \to H^{-1/2}(\Gamma)$$

are the Fredholm operators with zero indices. In addition, D and N are elliptic PDOs of orders -1 and $+1$, respectively.

By virtue of the regularity theorem, if the equations

$$D\varphi = f \tag{4.16}$$

and

$$N\psi = -g \tag{4.17}$$

have C^∞ right-hand sides and Γ is a C^∞ curve, then solutions φ and ψ on Γ belong to C^∞. Equations (4.16) and (4.17) or (4.14) and (4.15) are uniquely solvable for all k: $\Im k \geq 0$, $k \neq 0$ except for a discrete set of points. According to the definitions of Section 1.6, these points are the CNs and they are situated on the real axis with the only possible accumulation point at infinity. For equations (4.16) and (4.17), all CNs are isolated and have finite algebraic multiplicity. We will denote by $\sigma(D)$ and $\sigma(N)$ the sets of CNs for equations (4.16) and (4.17), respectively.

When the Dirichlet or the Neumann problem is solved for $k \in \sigma(D)$ or $k \in \sigma(N)$, (4.16) or (4.17) should be replaced by modified equations that are uniquely solvable for $k \in \sigma(D)$ or $k \in \sigma(N)$. Several methods of obtaining such equations are described in [62, 67]. Thus, for all values k such that $\Im k \geq 0$ and $k \neq 0$, the Dirichlet and the Neumann problems (4.1)–(4.5) are uniquely solvable. However, the solutions can be represented in the form of potentials (4.12) and (4.13) only if $k \notin \sigma(D)$ for the Dirichlet problem and $k \notin \sigma(N)$ for the Neumann problem. We have proved the following statement.

Theorem 34 *If $k \notin \sigma(D)$, then the operator*

$$D : H^{-1/2}(\Gamma) \to H^{1/2}(\Gamma)$$

is bounded and has a bounded inverse. In this case, equation (4.16) is uniquely solvable. If $k \notin \sigma(N)$, then the operator

$$N : H^{1/2}(\Gamma) \to H^{-1/2}(\Gamma)$$

is bounded and has a bounded inverse. In this case, equation (4.17) is uniquely solvable.

Assume that all Γ_j are unclosed smooth curves. Before to proceed to the analysis of the general case of the diffraction problem on Γ, we consider a particular case of intervals situated on one line, when

$$\Gamma = \{\mathbf{x} : y = 0, a_{2j-1} < x < a_{2j}, j = 1, \dots, J; a_i < a_j (i < j)\}.$$

In this case, D and N are convolution-type operators on the line R^1. Calculating the Fourier transforms of the kernels of these operators and using the convolution theorem and the properties of the Fourier transform, we rewrite equations (4.14) and (4.15) as

$$D\varphi \equiv \int_{-\infty}^{+\infty} \frac{1}{\sqrt{\xi^2 - k^2}} e^{ix\xi} \widehat{\varphi}(\xi)\, d\xi = f(x), \quad x \in \Gamma, \tag{4.18}$$

and

$$N\psi \equiv \int_{-\infty}^{+\infty} \sqrt{\xi^2 - k^2} e^{ix\xi} \widehat{\psi}(\xi)\, d\xi = g(x), \quad x \in \Gamma. \tag{4.19}$$

Here, $\widehat{\varphi}$ and $\widehat{\psi}$ denote the Fourier transforms of φ and ψ. We will look for solutions φ and ψ in the spaces of generalized functions (distributions) $\varphi \in \tilde{H}^{-1/2}(\Gamma)$ and $\psi \in \tilde{H}^{1/2}(\Gamma)$. In the case under consideration, these spaces can be described in terms of the Fourier transform:

$$\tilde{H}^s(\Gamma) = \left\{u : (1 + |\xi|)^s \widehat{u}(\xi) \in L_2(\mathrm{R}^1), \operatorname{supp} u \subset \overline{\Gamma}\right\}.$$

The left-hand sides of equations (4.18) and (4.19) define the boundary integrodifferential operators

$$D : \tilde{H}^{-1/2}(\Gamma) \to H^{1/2}(\Gamma)$$

and

$$N : \tilde{H}^{1/2}(\Gamma) \to H^{-1/2}(\Gamma),$$

where $H^s(\Gamma)$ is a restriction of $H^s(\mathrm{R}^1)$ on Γ. We have

$$\begin{aligned} (\xi^2 - k^2)^{-1/2} &= |\xi|^{-1} + O(|\xi|^{-3}), \\ (\xi^2 - k^2)^{1/2} &= |\xi| + O(|\xi|^{-1}) \end{aligned}$$

for $|\xi| \to \infty$; therefore, according to [66], D and N are Fredholm PDOs with the zero index and orders -1 and $+1$, respectively. These operators are invertible by virtue of the estimates

$$\left|(D\varphi, \varphi)_{L_2}\right| \geq \frac{1}{\sqrt{2}} \int_{-\infty}^{+\infty} \left|\xi^2 - k^2\right|^{-1/2} |\widehat{\varphi}(\xi)|^2\, d\xi > 0, \quad \varphi \neq 0,$$

and

$$\left|(N\psi,\psi)_{L_2}\right| \geq \frac{1}{\sqrt{2}} \int_{-\infty}^{+\infty} \left|\xi^2 - k^2\right|^{1/2} |\hat{\psi}(\xi)|^2 \, d\xi > 0, \quad \psi \neq 0.$$

These estimates prove that the kernels of these operators contain only the zero element: $\ker D = \{0\}$ and $\ker N = \{0\}$. Thus, by the Fredholm alternative, equations (4.18) and (4.19) or (4.14) and (4.15) are uniquely solvable.

Consider the general case of a system of curves of arbitrary shape Γ. The operators

$$D : \tilde{H}^{-1/2}(\Gamma) \to H^{1/2}(\Gamma)$$

and

$$N : \tilde{H}^{1/2}(\Gamma) \to H^{-1/2}(\Gamma),$$

are bounded, and their Fredholm property follows from the Gårding inequality [68]

$$\begin{aligned} \Re\langle (D + K_D)\varphi, \varphi\rangle &\geq \lambda_D \|\varphi\|^2_{-1/2}, \quad \forall\varphi \in \tilde{H}^{-1/2}(\Gamma), \\ \Re\langle (N + K_N)\psi, \psi\rangle &\geq \lambda_N \|\varphi\|^2_{1/2}, \quad \forall\psi \in \tilde{H}^{1/2}(\Gamma) \end{aligned}$$

with certain compact operators

$$\begin{aligned} K_D : &\ \tilde{H}^{-1/2}(\Gamma) \to H^{1/2}(\Gamma), \\ K_N : &\ \tilde{H}^{1/2}(\Gamma) \to H^{-1/2}(\Gamma). \end{aligned}$$

Equations (4.14) and (4.15) are uniquely solvable for arbitrary right-hand sides $f \in H^{1/2}(\Gamma)$ or $g \in H^{-1/2}(\Gamma)$, since the corresponding homogeneous equations have only trivial solutions. Indeed, if we assume that φ and ψ are nontrivial solutions of (4.14) and (4.15) for $f = 0$ and $g = 0$, then formulas (4.12) and (4.13) yield nontrivial solutions of homogeneous problems (4.1)–(4.5), which contradicts the uniqueness theorem. Thus, in the case of unclosed curves, equations (4.16) and (4.17) has a unique solution for all k: $\Im k \geq 0$, $k \neq 0$.

Theorem 35 *If $\Im k \geq 0$ and $k \neq 0$, then each operator*

$$D : \tilde{H}^{-1/2}(\Gamma) \to H^{1/2}(\Gamma)$$

and

$$N : \tilde{H}^{1/2}(\Gamma) \to H^{-1/2}(\Gamma)$$

is bounded and has a bounded inverse. Equations (4.16) and (4.17) are uniquely solvable.

When the diffraction problems are considered, it is important to investigate the asymptotic behaviour of solutions in the vicinities of the endpoints of $\partial\Gamma$. Let f and g be smooth functions, e.g., $f, g \in C^\infty(\overline{\Gamma})$. Applying the known results concerning the regularity of solutions of elliptic equations, one can easily show that φ and ψ are smooth functions on Γ (they belong to C^∞ if all $\Gamma_j \in C^\infty$) and have singularities

$$\varphi = O(\rho^{-1/2}), \quad \rho \to 0, \tag{4.20}$$

$$\psi = O(\rho^{1/2}), \quad \rho \to 0, \tag{4.21}$$

in the vicinities of the endpoints of $\partial\Gamma$ (ρ is the distance to an endpoint of the curve); note that estimates (4.20) and (4.21) are accurate with respect to the order.

4.3 Canonical problems: diffraction by strips and slots

4.3.1 Diffraction by a slot

Assume that a perfectly conducting plane with the slot $\mathrm{S} = \{(x, y) : -l < x < l, y = 0\}$ separates two media with the relative permittivities $\varepsilon_1 = 1$ and $\varepsilon_2 = \varepsilon$ (Fig. 4.1), and the longitudinal component of the incident plane wave is given by the formula

$$H_z^0(x, y) = A_h e^{ik_1(x \sin\theta - y \cos\theta)}.$$

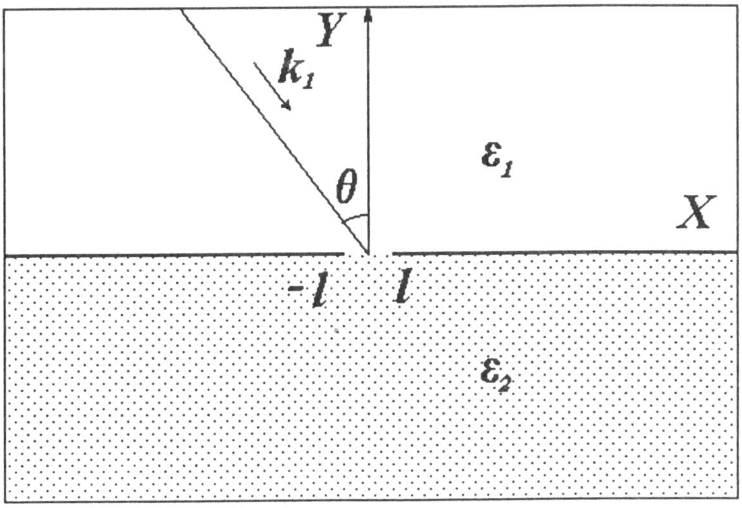

Figure 4.1

The longitudinal component of the diffracted field in the upper half-space can be represented as the sum

$$H_z^d = H_z^* + H_z^1,$$

where

$$H_z^*(x, y) = A_h e^{ik_1(x \sin\theta + y \cos\theta)} \tag{4.22}$$

is the longitudinal component of the field reflected from the plane without the slot, and H_z^1 is the perturbation caused by the slot. The longitudinal component of the field in the lower half-space (the transmitted field) is denoted by H_z^2. It is required to determine the longitudinal components of fields H_z^1 and H_z^2.

The solution to this problem is known in the so-called long-wave and short-wave approximations, i.e., when $k_1 l << 1$ and $k_1 l >> 1$ (see, for example, [22]). We will demonstrate how the long-wave approximation can be obtained by the method of approximate inversion.

The problem under investigation can be reduced to the boundary value problem (4.6), (4.3), (4.5) and (4.9) (problem *Hi*) with respect to function u defined as

$$u(x,y)=\begin{cases} H_z^1(x,y), & \text{for } y>0,\\ H_z^2(x,y), & \text{for } y<0.\end{cases}$$

The solution can be determined in the form of the Green's potentials

$$\begin{aligned} H_z^1(x_0,y_0) &= -\int_{-l}^{l} G_1(x_0,y_0,x,0;k_1)\frac{\partial H_z^1}{\partial y}\bigg|_S(x)dx, \quad y_0>0,\\ H_z^2(x_0,y_0) &= \int_{-l}^{l} G_2(x_0,y_0,x,0;k_1\sqrt{\varepsilon})\frac{\partial H_z^2}{\partial y}\bigg|_S(x)dx, \quad y_0<0. \end{aligned} \tag{4.23}$$

where G_1 and G_2 are the Green's functions in the halfplanes $y>0$ and $y<0$, respectively [formula (2.3.1)].

Satisfying conjugation conditions (4.7), we reduce problem *Hi* to the integral equation

$$\int_{-l}^{l}[\varepsilon_1 G_1(x_0,0,x,0,k_1)+\varepsilon_2 G_2(x_0,0,x,0,k_1\sqrt{\varepsilon})]\varphi(x)\,dx = f_H(x_0) \tag{4.24}$$

with respect to $\varphi(x)=\dfrac{\partial H_z^1}{\partial y}\bigg|_S(x)=-i\omega E_x^1(x,0)$, The right-hand side is determined as

$$f_h(x_0)=H_z^0(x_0,0)+H_z^*(x_0,0)=2H_z^0(x_0,0)=2A_h e^{ik_1x_0\sin\theta}. \tag{4.25}$$

The kernel of the operator in equation (4.24) is

$$K(x_0,x,\lambda)=\frac{i}{2}H_0^{(1)}(k_1|x_0-x|)+\varepsilon\frac{i}{2}H_0^{(1)}(k_1\sqrt{\varepsilon}|x_0-x|).$$

Assume that $l\to 0$. Then, as follows from representation (2.4) of the Green's function, the inverse operator can be represented by formula (1.134).

Let us analyze the first term $L_1^{-1}f_h$. It is given by formula (1.27) with

$$\alpha=1+\varepsilon,\quad f=f_h,$$

and

$$M_0(\lambda)=\frac{1+\varepsilon}{\pi}\ln\frac{2}{\gamma}+\left(\frac{1}{\pi}\ln\frac{1}{k_1}+\frac{\varepsilon}{\pi}\ln\frac{1}{k_1\sqrt{\varepsilon}}\right)+i\frac{1+\varepsilon}{2},$$

where $\gamma=1.780172418\ldots$ is the Euler constant.

Since f_h is an infinitely smooth function in the whole segment $[0,2\pi]$,

$$f_h(lt)=f_h(0)+f_h'(0)lt+O((lt)^2),\quad \text{as}\quad l\to 0,$$

and

$$L_1^{-1}f=O(\beta),\quad \text{as}\quad \beta\to 0,$$

where

$$\beta=\left(\frac{1}{\pi}\ln\frac{1}{l}\right)^{-1}.$$

Then, as follows from (1.134),

$$\tilde{K}^{-1}(k_1, l)f = L_1^{-1}(k_1, \beta)f + O(l^2), \quad \text{as} \quad l \to 0.$$

Within the error $O(l^2)$, we obtain

$$\varphi(x) = \frac{2A_h}{1+\varepsilon}\left(\frac{1}{\ln\dfrac{4}{l\gamma k_1} + \dfrac{\varepsilon}{2}\ln\dfrac{1}{\varepsilon} + i\dfrac{\pi}{2(1+\varepsilon)}} + ik_1 x\sin\theta + O(l^2)\right)\frac{1}{\sqrt{l^2 - x^2}} \quad \text{as} \quad l \to 0.$$

In a particular case of a slot in the screen in vacuum ($\varepsilon = 1$), the solution is simplified:

$$\varphi(x) = A_h\left(\frac{1}{\ln\dfrac{4}{l\gamma k_1} + i\dfrac{\pi}{4}} + ik_1 x\sin\theta + O(l^2)\right)\frac{1}{\sqrt{l^2 - x^2}} \quad \text{as} \quad l \to 0.$$

The field components H_z^1 and H_z^2 can be represented in the form of the Green's potentials (4.23).

4.3.2 Diffraction by a strip

Assume that the E-polarized plane wave

$$E_z^0(x, y) = A_e e^{ik_1(x\sin\theta - y\cos\theta)}$$

is incident on the perfectly conducting strip $S = \{(x, y) : -l < x < l, y = 0\}$ placed in a homogeneous medium.

Applying the approach set forth in the previous subsection, one can reduce the problem under investigation to boundary value problem *Ei*. The solution can be determined in the form of the SLP

$$E_z^1(x_0, y_0) = \frac{i}{4}\int_{-l}^{l} H_0^{(1)}(k\sqrt{(x_0 - x)^2 + y_0^2})\varphi(x)dx$$

with the unknown density $\varphi(x_0)$ satisfying the integral equation

$$\frac{i}{4}\int_{-l}^{l} H_0^{(1)}(k|x_0 - x|)\varphi(x)dx = f_e(x_0), \tag{4.26}$$

where $f_e(x) = -A_e e^{ik_1 x_0 \sin\theta}$.

In the long-wave and short-wave approximations, the solution to this equation is known. For sufficiently small l, we can obtain it by the method of appproximate inversion. Equation (4.26) coincides with that for the diffraction of a plane H-polarized wave by a slot, so we can immediately write the solution

$$\varphi(x) = -2A_e\left(\frac{1}{\ln\dfrac{4}{l\gamma k} + i\dfrac{\pi}{4}} + ik_1 x\sin\theta + O(l^2)\right)\frac{1}{\sqrt{l^2 - x^2}}, \quad \text{as} \quad l \to 0.$$

4.3.3 Diffraction by a screen with a rectangular slotted cavity

Assume that the longitudinal component of the field of the plane wave is given by the formula

$$H_z^0(x,y) = A_h e^{-ik_1(x\sin\theta - y\cos\theta)},$$

and the plane wave is incident on a planar screen with a rectangular slotted cavity (Fig.4.2); the boundary of the structure is perfectly conducting.

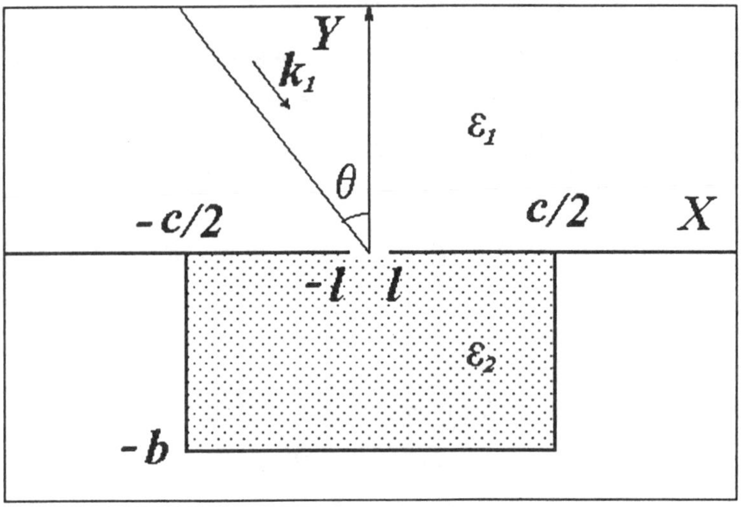

Figure 4.2

By analogy with Subsection 4.3.1, we introduce the (longitudinal components of) reflected field H_z^* (formula 4.22) and the sought-for secondary fields H_z^1 and H_z^2. The latter are determined in the form of the Green's potentials (4.23), where G_1 and G_2 are the Green's functions for the halfplane $I = \{(x,y) : y > 0\}$ and rectangle $II = \{(x,y) : -c/2 < x < c/2, -b < y < 0\}$. Satisfying conjugation conditions (4.7), we obtain the integral equation

$$\int_{-l}^{l} [\varepsilon_1 G_1(x_0, 0, x, 0; k_1) + \varepsilon_2 G_2(x_0, 0, x, 0; k_2)]\varphi(x)\,dx = f_h(x_0) \qquad (4.27)$$

with respect to $\varphi(x) = \dfrac{1}{\varepsilon_1}\dfrac{\partial H_z^1}{\partial y}\Big|_S(x)$, where the right-hand side is defined by formula (4.25).

Theorem 36 *For any real k_1, there exists the unique solution to equation (4.27). For sufficiently small l, this solution is given by the formula*

$$\varphi(x) = \frac{1}{(\varepsilon_1+\varepsilon_2)}\left(\frac{f_h(0)}{\ln 2 + \pi\left(\dfrac{1}{\beta} + \dfrac{M_0(\lambda)}{\varepsilon_1+\varepsilon_2}\right)} + f_h'(0)x + O(l^2)\right)\frac{1}{\sqrt{l^2-x^2}}, \quad as \quad l \to 0, \qquad (4.28)$$

where $\lambda = k_1^2/\varepsilon_1$ *and* $M_0(\lambda)$ *is expressed by (4.45).*

□ The existence and uniqueness follow from Theorem 38. Formula (4.28) is proved in Section 1.9. In particular, for small l, the existence and uniqueness immediately follow from formula (4.28), because the denominator

$$D(\lambda) = 1 + \frac{\pi}{\ln 2}\left(\frac{1}{\beta} + \frac{M_0(\lambda)}{\varepsilon_1 + \varepsilon_2}\right)$$

does not vanish at real values of λ. In fact, for sufficiently small β,

$$|\Im D(\lambda)| > \frac{\pi}{2\ln 2(\varepsilon_1 + \varepsilon_2)}.$$

Although the CNs cannot lie on the real axis, violating thus the existence and uniqueness, they can be very close to it and strongly influence the properties of the solution. This influence can be estimated in terms of the absolute value of $D(\lambda)$. Since, to within the terms of an order $O(\beta^3)$, $\frac{d\Im D(\lambda)}{d\lambda} = 0$, the minima of $|\Im D(\lambda)|$ are specified by the condition

$$\Re D(\lambda) = 0.$$

In the vicinity of a CN λ^*_{NM} (or pole λ_{NM}), the root $\tilde{\lambda}$ of this equation is given by the formula

$$\tilde{\lambda} = \Re\lambda^*_{NM} = \lambda_{NM} + \beta\frac{4}{bc}\frac{\delta_M}{\varepsilon_1 + \varepsilon_2} - \beta^2\frac{4}{bc}\frac{\delta_M}{\varepsilon_1 + \varepsilon_2}\Big(\frac{\Re C_{NM}(\lambda_{NM})}{\varepsilon_1 + \varepsilon_2} + \frac{\ln 2}{\pi}\Big), \quad \text{as} \quad \beta \to 0.$$

Let us consider two cases. In the first (resonant) case, we assume that the condition $\lambda \approx \Re\lambda^*_{NM}$ (i.e., $k_1 \approx \sqrt{\lambda^*_{NM}\varepsilon_1}$) is fulfilled for some N and M. In the second (nonresonant) case, λ is sufficiently distant from any CN.

In the nonresonant case, we use formula (1.134) to obtain the approximate solution of equation (4.27):

$$\varphi(x) = 2A_h\Big(\frac{\beta}{\pi(\varepsilon_1 + \varepsilon_2)} + O(\beta^2)\Big)\frac{1}{\sqrt{l^2 - x^2}}, \quad \text{as} \quad \beta \to 0. \tag{4.29}$$

In the resonant case (exactly at $k_1 = \sqrt{\lambda^*_{NM}\varepsilon_1}$), the formula becomes substantially different:

$$\varphi(x) = -2A_h\Big(\frac{2i}{\pi\varepsilon_1} + O(l)\Big)\frac{1}{\sqrt{l^2 - x^2}}, \quad \text{as} \quad l \to 0. \tag{4.30}$$

Substituting (4.29) into (4.23), we obtain

$$H_z^1(x,y) = -i\beta\frac{A_h\varepsilon_1}{\pi(\varepsilon_1 + \varepsilon_2)}\int_{-1}^{1} H_0^{(1)}(k_1\sqrt{(x - lt)^2 + y^2})\frac{dt}{\sqrt{1 - t^2}} + O(\beta^2), \quad \text{as} \quad \beta \to 0.$$

In the resonant case,

$$H_z^1(x,y) = -\frac{2A_h}{\pi}\int_{-1}^{1} H_0^{(1)}(k_1\sqrt{(x - lt)^2 + y^2})\frac{dt}{\sqrt{1 - t^2}} + O(l^2), \quad \text{as} \quad l \to 0. \tag{4.31}$$

The fields obtained in these two cases are substantially different. Thus, for large $r = \sqrt{x^2 + y^2}$, in the nonresonant case, we have

$$H_z^1(x,y) = \beta A_h\frac{\varepsilon_1}{\varepsilon_1 + \varepsilon_2}\sqrt{\frac{2}{\pi k_1 r}}e^{i(k_1 r - \frac{3\pi}{4})} \quad \text{for} \quad r \gg 1,$$

and the amplitude of the field perturbed by the slot falls as the slotwidth decreases.

In the resonant case,

$$H_z^1(x,y) = 2A_h\sqrt{\frac{2}{\pi k_1 r}}e^{i(k_1 r+\frac{3\pi}{4})} \quad \text{for} \quad r \gg 1,$$

and the amplitude of field $H_z^1(x,y)$ has the same order as that of the incident field, regardless of the size of the slot, while the phase shift equals $3\pi/2$.

Inside the rectangular cavity, the field is also excited. In the nonresonant case, it is expressed by the formula

$$H_z^2(x,y) = \frac{2A_h\beta\varepsilon_2}{\pi(\varepsilon_1+\varepsilon_2)}\int_{-1}^{1} G_2(x,y,lt,0,k_2)\frac{dt}{\sqrt{1-t^2}} + O(\beta^2) \quad \text{as} \quad \beta\to 0, \tag{4.32}$$

and this field is obviously small for small l.

In the resonant case,

$$\begin{aligned} H_z^2(x,y) &= \frac{4iA_h(\varepsilon_1+\varepsilon_2)}{\beta\varepsilon_1}\cos\frac{\pi N x}{c}\cos\frac{\pi M y}{b} \\ -\frac{4iA_h\varepsilon_2}{\pi\varepsilon_1}&\int_{-1}^{1}\tilde{G}_2(x,y,lt,0;k_2)\frac{dt}{\sqrt{1-t^2}} + O(l^2), \quad \text{as} \quad l\to 0, \end{aligned} \tag{4.33}$$

where

$$\tilde{G}_2(x,y,lx_0,0;k_2) = G_2(x,y,lx_0,0;k_2) - \frac{4\delta_N\delta_M}{bc\varepsilon_2}\frac{\cos\frac{\pi N x}{c}\cos\frac{\pi N l x_0}{c}\cos\frac{\pi M y}{b}}{\lambda_{NM}-\lambda}.$$

Thus, when slotwidth l tends to zero, field $H_z^2(x,y)$ in the rectangular cavity is an unbounded quantity.

Formulas (4.30)–(4.33) were applied for calculating the amplitude and phase distributions of the scattered fileds in the nonresonant and resonant cases. The results obtained for the oblique incidence of the plane unit-amplitude H-polarized wave incident at the angle $\theta = \pi/4$ are presented as examples. Figures 4.3–4.6 demonstrate the amplitude and phase (over the period interval 2π) distributions calculated in the nonresonant (Figs. 4.3 and 4.5) and resonant (Figs. 4.4 and 4.6) cases, respectively (in the latter case, $\lambda \approx \lambda_{2,1}$). We see that in the nonresonant case, the perturbation caused by the slot is small, while in the resonant case, it strongly influences the amplitude and phase distributions.

4.3.4 Scattering by a circular slotted cylinder

Let a plane wave with the longitudinal field component

$$H_z^0(x,y) = A_h e^{-ik_1(x\cos\theta + y\sin\theta)}$$

be incident on a perfectly conducting circular slotted cylinder (Fig. 4.7).

Represent the longitudinal component of the scattered fileld as the sum

$$H_z^{sc} = H_z^* + H_z^1,$$

where H_z^* is longitudinal component of the field scattered by the perfectly conducting cylinder without the slot and H_z^1 is the perturbation caused by the slot. Denote the longitudinal component of the field inside the resonator by H_z^2.

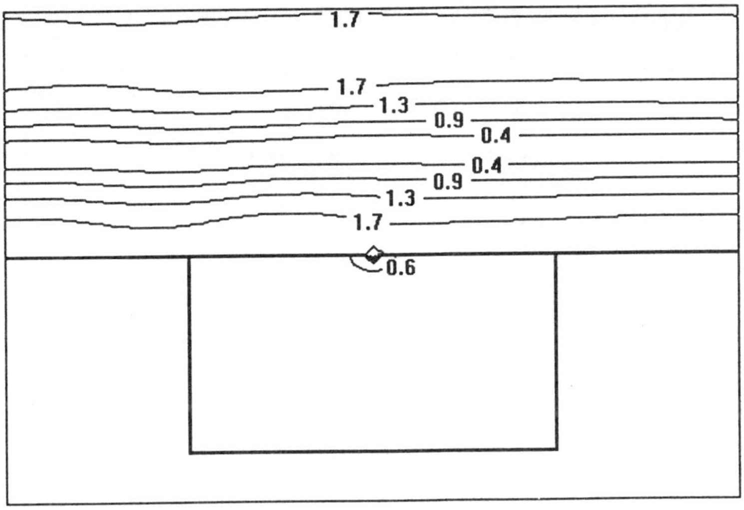

Figure 4.3

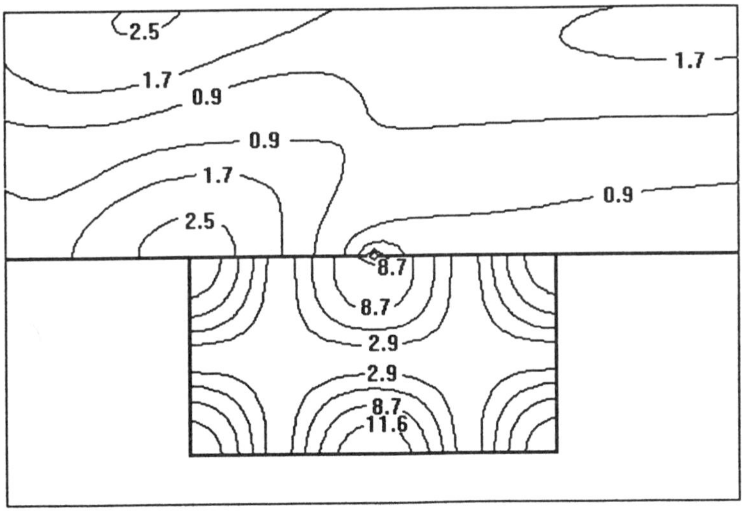

Figure 4.4

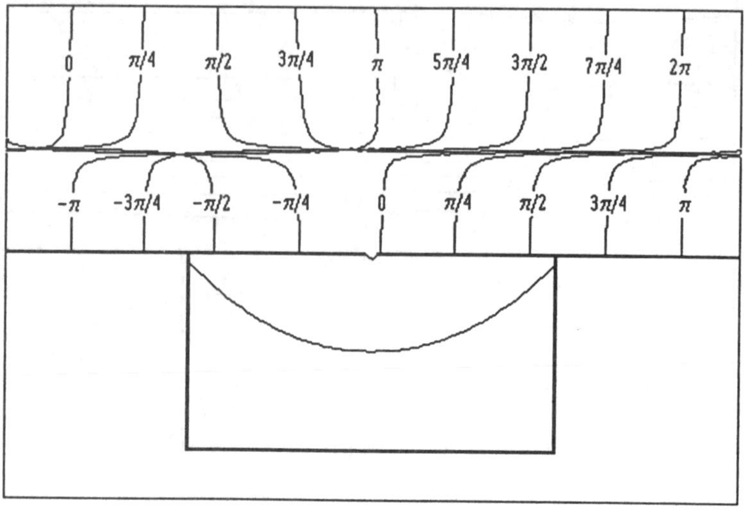

Figure 4.5

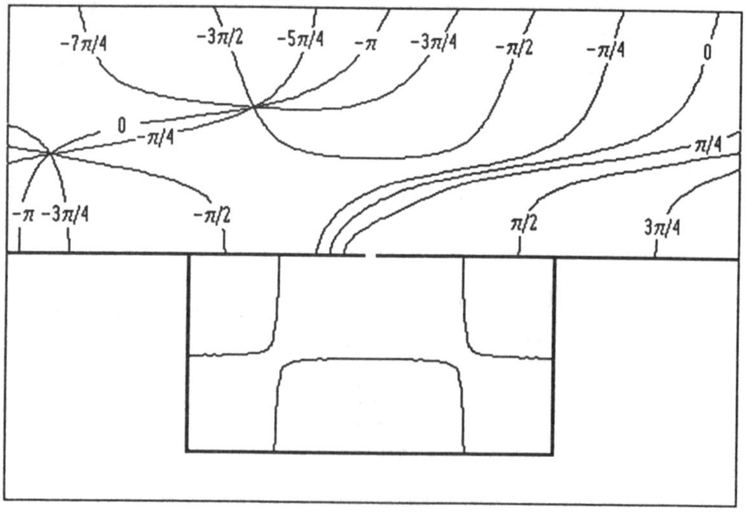

Figure 4.6

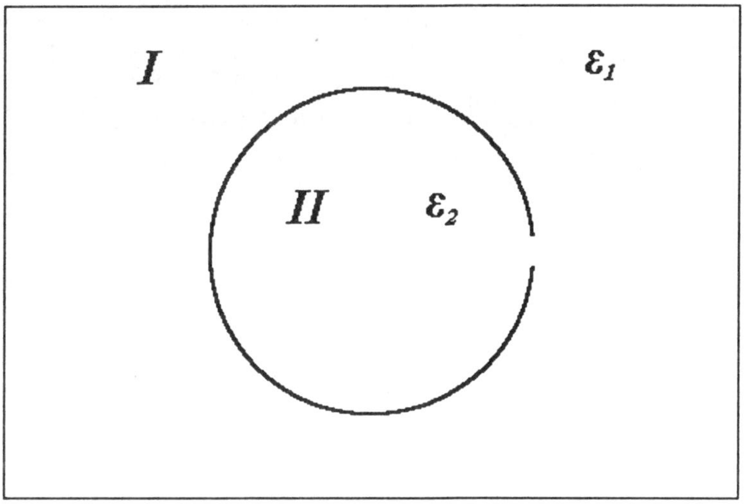

Figure 4.7

Longitudinal components H_z^1 and H_z^2 of the secondary fields (as in Subsection 4.3.1, they can be specified as solutions to the corresponding boundary value proble *Hi*) are sought in the form of Green's potentials (4.41), and the problem is reduced to the integral equation:

$$R\int_{-l}^{l}[\varepsilon_1 G_1(R,\phi_0,R,\phi;k_1)+\varepsilon_2 G_2(R,\phi_0,R,\phi;k_2)]\varphi(\phi)\,d\phi = f_h(\phi_0), \quad -l<\phi_0<l, \tag{4.34}$$

with respect to the unknown function $\varphi(\phi)=\left.\frac{1}{\varepsilon_1}\frac{\partial H_z^1}{\partial r}\right|_{r=R}(\phi)$. The operator in (4.34) is the same as in formula (4.40), and the right-hand side is

$$f_h(\phi_0)=H_z^0(R,\phi_0)+H_z^*(R,\phi_0), \quad -l<\phi_0<l.$$

In order to find field $H_z^*(R,\phi_0)$, we must solve the problem of scattering by a circular perfectly conducting cylinder, which is one of the canonical problems of electromagnetics [22]. This field can be expressed by the Green's formula

$$H_z^*(r_0,\phi_0)=-R\int_0^{2\pi}G_1(r_0,\phi_0,R,\phi;k_1)\left.\frac{\partial H_z^*}{\partial r}\right|_{r=R}(\phi)d\phi,$$

where G_1 is the Green's function of the exterior of the circle given by formula (2.9). After some calculations, we obtain

$$H_z^*(r_0,\phi_0)=-\sum_{n=-\infty}^{+\infty}\frac{H_n^{(1)}(k_1r_0)J_n'(k_1R)}{H_n^{(1)\prime}(k_1R)}e^{in(\theta-\phi_0-\frac{\pi}{2})}. \tag{4.35}$$

By analogy with Theorem 36, the following statement can be formulated and proved.

Theorem 37 *For any real λ, there exists the unique solution to equation (4.34). For sufficiently small l, this solution is given by the formula*

$$\varphi(\phi)=\frac{1}{(\varepsilon_1+\varepsilon_2)}\left(\frac{f_h(0)}{\ln 2+\pi\left(\frac{1}{\beta}+\frac{M_0(\lambda)}{\varepsilon_1+\varepsilon_2}\right)}+f_h'(0)\phi+O(l^2)\right)\frac{1}{\sqrt{l^2-\phi^2}},\quad as\quad l\to 0,$$

where $M_0(\lambda)$ is expressed by (4.45).

Fields H_z^1 and H_z^2 outside and inside the cylinder are expressed by the Green's formulas similar to (4.23). As in the previous case, two cases of scattering can be considered, resonant and nonresonant, depending on the closeness of real spectral parameter λ to complex CNs λ_{NM}. We will demonstrate the difference between these two cases by constructing far field patterns $\Phi(\phi)$ of the total scattered field H_z^{sc}.

In the nonresonant case,

$$\Phi(\phi)=-\sqrt{\frac{2}{\pi}}\sum_{n=-\infty}^{+\infty}\frac{e^{-in(\phi+\frac{\pi}{2})}}{H_n^{(1)\prime}(k_1R)}[J_n'(k_1R)e^{in(\theta-\frac{\pi}{2})}+\beta\frac{\varepsilon_1 f_h(0)}{2\pi k_1R(\varepsilon_1+\varepsilon_2)}],\tag{4.36}$$

where, as follows from formulas (4.35) and (4.34),

$$f_h(0)=e^{-ik_1R\cos\theta}-\sum_{n=-\infty}^{+\infty}\frac{H_n^{(1)}(k_1R)J_n'(k_1R)}{H_n^{(1)\prime}(k_1R)}e^{in(\theta-\frac{\pi}{2})}.\tag{4.37}$$

In the resonant case, the far-field patterns differ substantially:

$$\Phi(\phi)=-\sqrt{\frac{2}{\pi}}\sum_{n=-\infty}^{+\infty}\frac{e^{-in(\varphi+\frac{\pi}{2})}}{H_n^{(1)\prime}(k_1R)}\left[J_n'(k_1R)e^{in(\theta-\frac{\pi}{2})}-i\frac{\varepsilon_1 f_h(0)}{2\pi k_1RC_h}\right],\tag{4.38}$$

where

$$C_h=\frac{2\varepsilon_1}{\pi^2(k_1R)^2}\sum_{n=0}^{\infty}\frac{\delta_n}{(J_n'(k_1R))^2+(N_n'(k_1R))^2}.\tag{4.39}$$

Examples of the far field patterns in the resonant and nonresonant cases calculated by formulas (4.36)–(4.39) are shown in Figs.4.8 and 4.9.

Figure 4.8 demonstrated the far field pattern in the resonant case when $k_1\approx j_{0,3}'\sqrt{\frac{\varepsilon_1}{\varepsilon_2}}$. The corresponding far field pattern of a completely shielded pivot (perfectly conducting cylinder) calculated for $k_1=j_{0,3}'\sqrt{\frac{\varepsilon_1}{\varepsilon_2}}$, which may be considered as the limiting case of the nonresonant scattering, is shown in Fig. 4.9.

4.4 Eigenoscillations of open and closed slot resonators

4.4.1 Closed rectangular slot resonator

The problem on eigenfrequencies of a closed rectangular slot resonator (the cylindrical structure with the cross section shown in Fig. 4.10) can be formulated as the boundary

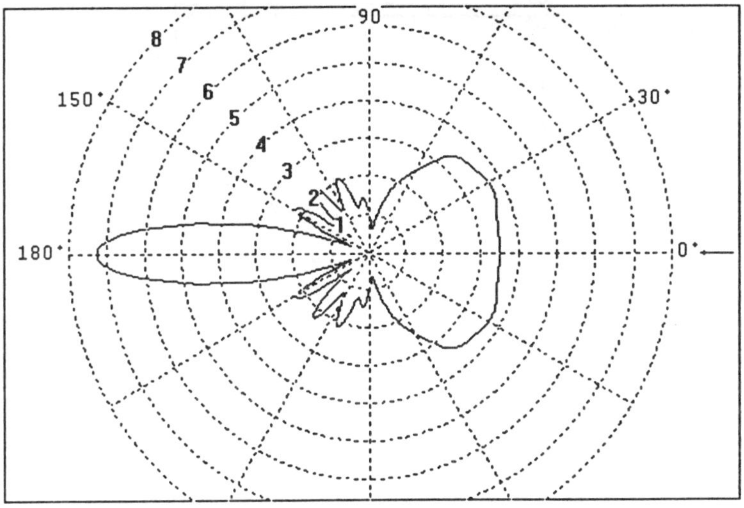

Figure 4.8

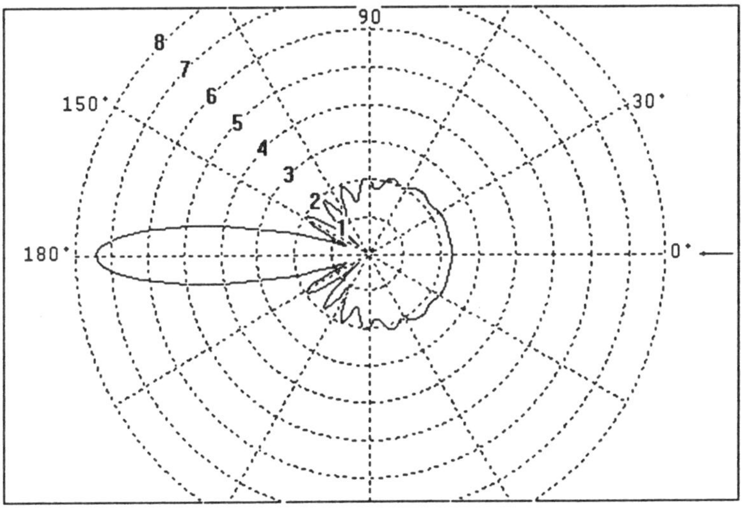

Figure 4.9

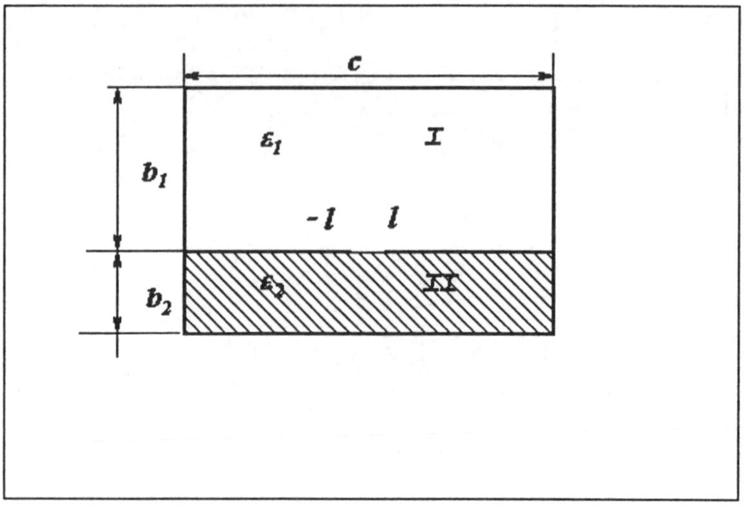

Figure 4.10

value problem (4.6), (4.3), (4.5) and (4.7) (problem *H*). Its solutions can be determined in the form of Green's potentials (4.23), where G_1 and G_2 are the Green's functions for rectangles I and II.

Satisfying the conjugation conditions (4.7) on the slot Γ, we reduce the problem to the homogeneous integral equation

$$\int_{-l}^{l} [\varepsilon_1 G_1(x_0, 0, x, 0; k_1) + \varepsilon_2 G_2(x_0, 0, x, 0, k_2)]\varphi(x)\, dx = 0,$$

The kernel of the integral operator has the form

$$K(x_0, x; \lambda) = \frac{4}{c}\sum_{n=0}^{\infty}\left[\frac{1}{b_1}\sum_{m=0}^{\infty}\frac{\delta_n\delta_m}{\lambda_{nm}^{(1)} - \lambda} + \frac{1}{b_2}\sum_{m=0}^{\infty}\frac{\delta_n\delta_m}{\lambda_{nm}^{(2)} - \lambda}\right]\varphi_n(x)\varphi_n(x_0),$$

where

$$\varphi_n(x) = \begin{cases} \cos\dfrac{\pi n x}{c}; & n = 0, 2, \dots, \\ \sin\dfrac{\pi n x}{c}; & n = 1, 3, \dots, \end{cases}$$

$\lambda = \omega^2\varepsilon_0\mu_0$ is the spectral parameter, and

$$\lambda_{nm}^{j} = \frac{\pi^2}{\varepsilon_j}\left(\frac{n^2}{c^2} + \frac{m^2}{b_j^2}\right), \quad n, m = 0, 1, 2, \dots, \quad j = 1, 2,$$

are the poles of the Green's functions. It is necessary to determined CNs of operator-valued function $K(\lambda)$, which correspond to the eigenfrequencies of the resonator under study.

Evidently, $K(\lambda)$ can be represented in the form (1.127) with the coefficients

$$\beta_n(\lambda) = \frac{4}{cb_1}\sum_{m=0}^{\infty}\frac{\delta_m}{\lambda_{nm}^{(1)}-\lambda} + \frac{4}{cb_2}\sum_{m=0}^{\infty}\frac{\delta_m}{\lambda_{nm}^{(2)}-\lambda}, \qquad n = 0,1,2,\ldots.$$

Then, the following theorem can be formulated.

Theorem 38 *The spectrum of operator-valued function $K(\lambda)$ consists of a countable set of real CNs. For any j, n, m_1, and $m_2 \neq m_1$, there exists at least one CN between two poles $\lambda_{nm_1}^{(j)}$ and $\lambda_{nm_2}^{(j)}$.*

□ The system of trigonometric functions $\{\varphi_n\}_{n=0}^{\infty}$ is complete in $L^1(-\pi/2,\pi/2)$; therefore, it is complete in $L^1(-l,l)$. For $\lambda < 0$, $\beta_n(\lambda) > 0$ for all n, and, for any real λ, $\beta_n'(\lambda) > 0$ for all n. Thus, coefficients $\beta_n(\lambda)$ satisfy the conditions of Theorem 18. □

For sufficiently small l, the CNs can be obtained by the method of local semi-inversion. Choosing a pole $\lambda_{NM}^{(j)}$ with even N, we can directly apply formula (1.134), expressing $M_0(\lambda)$, $C_{NM}(\lambda)$, α, μ_{NM}, and c_N explicitly (some of them are supplied with additional index j):

$$M_0(\lambda) = -\frac{\varepsilon_1+\varepsilon_2}{\pi}\frac{\ln\pi}{c} - \frac{\varepsilon_1+\varepsilon_2}{\pi}\ln 2 + \frac{\varepsilon_1}{\pi}\sum_{j=1}^{2}\sum_{n=1}^{\infty}\frac{e^{-\frac{2\pi n b_j}{c}}}{n\sinh\frac{2\pi n b_j}{c}}$$

$$+\frac{4}{c}\sum_{j=1}^{2}\sum_{n=1}^{\infty}\sum_{m=0}^{\infty}\frac{\delta_m}{b_j}\left(\frac{1}{\lambda_{2nm}-\lambda}-\frac{1}{\lambda_{2nm}}\right) + \frac{2}{b_j c}\sum_{m=0}^{\infty}\frac{\delta_m}{\lambda_{0m}-\lambda},$$

$$\mu_{NM}^{j} = -\frac{4\delta_N\delta_M}{b_j c}, \qquad c_N = 1,$$

$$C_{NM}^{j} = M_0(\lambda) - \frac{4}{b_j c}\frac{\delta_N\delta_M}{\lambda_{NM}-\lambda}.$$

For even N, according to formula (1.143), CN λ_{NM}^{*} is expressed as

$$\lambda_{NM}^{*} = \lambda_{NM} + \beta\frac{4}{b_j c}\frac{\delta_N\delta_M}{\varepsilon_1+\varepsilon_2} - \beta^2\frac{4}{b_j c}\frac{\delta_N\delta_M}{\varepsilon_1+\varepsilon_2}\left(\frac{C_{NM}(\lambda_{NM})}{\varepsilon_1+\varepsilon_2} + \frac{\ln 2}{\pi}\right) + O(\beta^3) \quad \text{as} \quad \beta \to 0.$$

For odd N, the CNs are given by formula (1.148), i.e.,

$$\lambda_{NM}^{*} = \lambda_{NM} + l^2\frac{2\delta_M\pi^3N^2}{cb_j^3(\varepsilon_1+\varepsilon_2)} + O(l^4\ln l) \quad \text{as} \quad l \to 0.$$

4.4.2 Open rectangular slot resonator

As in the previous subsection, the problem of eigenfrequencies of an open rectangular slot resonator (whose cross section is shown in Fig. 4.2) can be reduced to uniform integral equation (4.24), where G_1 and G_2 are the Green's functions of the halfplane $I = \{(x,y) : y > 0\}$ and rectangle $II = \{(x,y) : -c/2 < x < c/2,\ -b < y < 0\}$.

The kernel of the operator has the form

$$K(x_0,x;\lambda) = \varepsilon_1\frac{i}{2}H_0^{(1)}(\sqrt{\lambda\varepsilon_1}|x_0-x|) + \frac{4}{bc}\sum_{n=0}^{\infty}\sum_{m=0}^{\infty}\frac{\delta_n\delta_m}{\lambda_{nm}-\lambda}\varphi_n(x_0)\varphi_n(x), \qquad (4.40)$$

where

$$\lambda_{nm} = \frac{\pi^2}{\varepsilon_2}\left(\frac{n^2}{c^2} + \frac{m^2}{b^2}\right), \quad n, m = 0, 1, 2, \ldots,$$

and $\lambda = \omega^2 \varepsilon_0 \mu_0$ is the spectral parameter.

Theorem 39 *The spectrum of operator-valued function $K(\lambda)$ with kernel (4.40) is at most countable.*

□ Let us consider the sesquilinear form

$$F(\lambda, \varphi) = (K(\lambda)\varphi, \varphi).$$

We will prove that for any real $\lambda > 0$, the equality $F(\lambda, \varphi) = 0$ implies $\varphi = 0$. To this end, it would be sufficient to prove that $\Im F(\lambda, \varphi) > 0$ for any φ.

Evidently,

$$\Im F(\lambda, \varphi) = \int_{-l}^{l} dx_0 \int_{-l}^{l} J_0(k_1|x_0 - x|)\bar{\varphi}(x_0)\varphi(x)dx, \quad k_1 = \sqrt{\lambda \varepsilon_1}.$$

Then, taking into account the representation of the Bessel function [78], we obtain

$$\Im F(\lambda, \varphi) = \int_{-l}^{l} dx_0 \int_{-l}^{l} \left(\frac{1}{\pi}\int_0^{\pi} \cos(k_1|x_0 - x|\sin t)dt\right) \bar{\varphi}(x_0)\varphi(x)dx$$

$$= \frac{1}{\pi}\int_0^{\pi} dt \left|\int_{-l}^{l} \cos(k_1 x \sin t)\,\varphi(x)dx\right|^2 + \frac{1}{\pi}\int_0^{\pi} dt \left|\int_{-l}^{l} \sin(k_1 x \sin t)\,\varphi(x)dx\right|^2 \geq 0.$$

The equality to zero can be attained only for $\varphi(x) \equiv 0$ due to the completeness of the trigonometric system. Thus, the resolvent set of $K(\lambda)$ is not empty and, therefore, the spectrum is at most countable. □

For sufficiently small l, we will determine the CNs by the method of local semi-inversion. Choosing a pole λ_{NM} with even N, we obtain explicit expressions for quantities $M_0(\lambda)$, α, μ_{NM}, and c_N that enter formula (1.134):

$$M_0(\lambda) = \frac{\varepsilon_1}{2\pi}\ln\frac{1}{\lambda\varepsilon_1} + \frac{\varepsilon_1}{\pi}\ln\frac{2}{\gamma} + \frac{i\varepsilon_1}{2} - \frac{\varepsilon_2 \ln \pi}{\pi\ \ c} - \frac{\varepsilon_2}{\pi}\ln 2 + \frac{\varepsilon_2}{\pi}\sum_{n=1}^{\infty}\frac{e^{-\frac{2\pi n b}{c}}}{n \sinh\frac{2\pi n b}{c}}$$

$$+\frac{4}{bc}\sum_{n=1}^{\infty}\sum_{m=0}^{\infty}\delta_m\left(\frac{1}{\lambda_{2nm} - \lambda} - \frac{1}{\lambda_{2nm}}\right) + \frac{2}{bc}\sum_{m=0}^{\infty}\frac{\delta_m}{\lambda_{0m} - \lambda},$$

$$\mu_{NM} = -\frac{4}{bc}, \quad c_N = 1,$$

$$C_{NM}(\lambda) = M_0(\lambda) - \frac{4}{bc}\frac{\delta_N \delta_M}{\lambda_{NM} - \lambda}.$$

Since all poles λ_{NM} are real, the imaginary part of $C_{NM}(\lambda_{NM})$ can be separated; as a result, we obtain

$$\Im C_{NM}(\lambda_{NM}) = \frac{\varepsilon_1}{2}.$$

Then, applying formula (1.143), we derive the expression for the CNs in the vicinity of pole λ_{NM} with even N:

$$\lambda^*_{NM} = \lambda_{NM} + \beta\frac{4}{bc}\frac{\delta_N\delta_M}{\varepsilon_1+\varepsilon_2} - \beta^2\frac{4}{bc}\frac{\delta_N\delta_M}{\varepsilon_1+\varepsilon_2}\Big(\frac{\Re C_{NM}(\lambda_{NM})}{\varepsilon_1+\varepsilon_2} + \frac{\ln 2}{\pi}\Big)$$

$$-i\beta^2\frac{2}{bc}\frac{\delta_N\delta_M\varepsilon_1}{(\varepsilon_1+\varepsilon_2)^2} + O(\beta^3) \quad \text{as} \quad \beta \to 0.$$

As expected, the CN is complex with the negative imaginary part.

In the case of odd N, we should use formulas (1.145)-(1.147). As follows from the representation of kernel $K(x, x_0; \lambda)$ at real λ,

$$N(\lambda, l) = \nu(\lambda, l)P_2,$$

where

$$\nu(\lambda, l) = (\varepsilon_1^2 + \varepsilon_2^2)\frac{\lambda}{4\pi} - i\frac{1}{\ln l}\frac{\lambda\varepsilon_1^2}{8}.$$

Then, the CN is expressed by the formula

$$\lambda^*_{NM} = \lambda_{NM} + l^2\frac{2\delta_M\pi^3N^2}{bc^3(\varepsilon_1+\varepsilon_2)} + l^4\ln l\frac{\lambda_{NM}\delta_M\pi^3N^2}{2bc^3}\frac{\varepsilon_1^2+\varepsilon_2^2}{(\varepsilon_1+\varepsilon_2)^2}$$

$$-il^4\frac{\lambda_{NM}\delta_M}{4bc^3}\frac{\pi^4N^2\varepsilon_1^2}{(\varepsilon_1+\varepsilon_2)^2} + r(l),$$

where $r(l)$ is a complex quantity,

$$\Re r(l) = O(l^4) \quad \text{and} \quad \Im r(l) = O(l^6) \quad \text{as} \quad l \to 0.$$

We see that in this case, the CN also has the negative imaginary part.

4.4.3 Slotted resonator with circular cross section

The problem on eigenfrequencies of this type of resonator (shown in Fig. 4.7) also can be reduced to a homogeneous logarithmic integral equation. To this end, fields H_z^1 and H_z^2 outside and inside the resonator are sought as the Green's potentials

$$\begin{aligned} H_z^1(r_0,\phi_0) &= -R\int_{-l}^{l} G_1(r_0,\phi_0,R,\phi;k_1)\frac{\partial H_z^1}{\partial r}\Big|_S(\phi)d\phi, \quad r_0 > R, \\ H_z^2(r_0,\phi_0) &= R\int_{-l}^{l} G_2(r_0,\phi_0,R,\phi;k_1\sqrt{\varepsilon})\frac{\partial H_z^2}{\partial r}\Big|_S(\phi)d\phi, \quad r_0 < R, \end{aligned} \tag{4.41}$$

where G_1 and G_2 are the Green's functions for the exterior of the circle $r > R$ [formula (2.9)] and for the circle $r < R$ [formula (2.6)], respectively.

Satisfying conjugation conditions (4.7) we obtain the equation

$$\int_{-l}^{l}[\varepsilon_1 G_1(R,\phi_0,R,\phi;\lambda\sqrt{\varepsilon_1}) + \varepsilon_2 G_2(R,\phi_0,R,\phi;\lambda\sqrt{\varepsilon_2})]\varphi(\phi)\,d\phi = 0, \quad -l < \phi_0 < l,$$

with respect to $\varphi(\phi) = \frac{1}{\varepsilon_1}\frac{\partial H_z^1}{\partial r}\Big|_{r=R}(\phi)$, where $\lambda = \omega\sqrt{\varepsilon_0\mu_0}$ is the spectral parameter.

The kernel of the operator has the form

$$\begin{aligned} K(\phi_0,\phi;\lambda) = &-\frac{\varepsilon_1}{\pi\lambda\sqrt{\varepsilon_1}R}\sum_{n=0}^{\infty}\frac{\delta_n H_n^{(1)}(\lambda\sqrt{\varepsilon_1}R)}{H_n^{(1)\prime}(\lambda\sqrt{\varepsilon_1}R)}\cos n(\phi-\phi_0) \\ &+\frac{\varepsilon_2}{\pi\lambda\sqrt{\varepsilon_1}R}\sum_{n=0}^{\infty}\frac{\delta_n J_n(\lambda\sqrt{\varepsilon_1}R)}{J_n'(\lambda\sqrt{\varepsilon_2}R)}\cos n(\phi-\phi_0). \end{aligned} \tag{4.42}$$

Theorem 40 *The spectrum of operator-valued function $K(\lambda)$ with kernel (4.42) is at most countable.*

□ As in Theorem 39, it would be sufficient to prove that $\Im(K(\lambda)\varphi,\varphi) > 0$ for any real λ and any function $\varphi \neq 0$.

Separating the imaginary part of kernel (4.42), we obtain

$$\begin{aligned} \Im(K(\lambda)\varphi,\varphi) &= \int_{-l}^{l} d\phi_0 \int_{-l}^{l} \frac{1}{\pi^2(\lambda R)^2}\sum_{n=0}^{\infty}\frac{\delta_n \cos n(\phi-\phi_0)}{\left(J_n'(\lambda\sqrt{\varepsilon_1}R)\right)^2 + \left(N_n'(\lambda\sqrt{\varepsilon_1}R)\right)^2}\bar\varphi(\phi_0)\varphi(\phi)\,d\phi \\ &= \frac{1}{\pi^2(\lambda R)^2}\sum_{n=0}^{\infty}\frac{\delta_n}{\left(J_n'(\lambda\sqrt{\varepsilon_1}R)\right)^2 + \left(N_n'(\lambda\sqrt{\varepsilon_1}R)\right)^2} \\ &\times\left(\left|\int_{-l}^{l}\cos(n\phi)\varphi(\phi)\,d\phi\right|^2 + \left|\int_{-l}^{l}\sin(n\phi)\varphi(\phi)\,d\phi\right|^2\right) > 0. \end{aligned} \tag{4.43}$$

The obtained inequality proves the theorem. □

Making the change of variables (1.129), we introduce the transformed operator-valued function $\tilde K(\lambda,l)$ (see Section 1.9). Choosing a pole $\lambda_{NM} = \frac{j'_{N,M}}{\sqrt{\varepsilon_2}R}$, where $j'_{N,M}$ is the Mth root of the derivative of the Bessel function $J_N'(z)$, we can represent $\tilde K(\lambda,l)$ in the vicinity of λ_{NM} as

$$\tilde K(\lambda,l)\tilde\varphi = K_{NM}(\lambda,l)\tilde\varphi + \frac{\mu_{NM}}{\lambda-\lambda_{NM}}\left((\tilde\varphi,\varphi_N)\varphi_N + (\tilde\varphi,\psi_N)\psi_N\right), \tag{4.44}$$

where

$$\varphi_N(t) = \cos(Nlt), \qquad \psi_N(t) = \sin(Nlt),$$

and

$$\mu_{NM} = -\frac{\varepsilon_2}{\pi j'_{N,M}\sqrt{\varepsilon_2}R}\frac{1}{1-\frac{N^2}{j'^2_{N,M}}}. \tag{4.45}$$

Operator-valued function $K_{NM}(\lambda,l)$ is analytical at point λ_{NM}. In the given case, $M_0(\lambda)$ is given by the formula

$$M_0(\lambda) = \frac{\varepsilon_1}{2\pi\lambda\sqrt{\varepsilon_1}R}\frac{H_0^{(1)}(\lambda\sqrt{\varepsilon_1}R)}{H_1^{(1)}(\lambda\sqrt{\varepsilon_1}R)}$$

$$+\frac{\varepsilon_1\lambda\sqrt{\varepsilon_1}R}{\pi}\sum_{n=1}^{\infty}\frac{H_{n-1}^{(1)}(\lambda\sqrt{\varepsilon_1}R)}{H_n^{(1)}(\lambda\sqrt{\varepsilon_1}R)}\frac{1}{n\left(n-\lambda\sqrt{\varepsilon_1}R\frac{H_{n-1}^{(1)}(\lambda\sqrt{\varepsilon_1}R)}{H_n^{(1)}(\lambda\sqrt{\varepsilon_1}R)}\right)}$$

$$-\frac{\varepsilon_2}{2\pi\lambda\sqrt{\varepsilon_2}R}\frac{J_0(\lambda\sqrt{\varepsilon_2}R)}{J_1(\lambda\sqrt{\varepsilon_2}R)}+\frac{\varepsilon_2\lambda\sqrt{\varepsilon_2}R}{\pi}\sum_{n=1}^{\infty}\frac{J_{n+1}(\lambda\sqrt{\varepsilon_2}R)}{J_n(\lambda\sqrt{\varepsilon_2}R)}\frac{1}{n\left(n-\lambda\sqrt{\varepsilon_2}R\frac{J_{n+1}(\lambda\sqrt{\varepsilon_2}R)}{J_n(\lambda\sqrt{\varepsilon_2}R)}\right)}.$$

Function $C_{NM}(\lambda)$ is defined by formula (1.137) with μ_{NM} given by (4.45).

The imaginary part of $M_0(\lambda)$ and $C_{NM}(\lambda)$ at real λ can be separated; as a result, we obtain

$$\Im M_0(\lambda)=\Im C_{NM}(\lambda)$$

$$=\frac{2}{\pi^2(\lambda R)^2}\sum_{n=0}^{\infty}\frac{\delta_n}{\left(J_n'(\lambda\sqrt{\varepsilon_1}R)\right)^2+\left(N_n'(\lambda\sqrt{\varepsilon_1}R)\right)^2}+O(l^2)\quad\text{as}\quad l\to 0.$$

Obviously, the pole pencil in representation (4.44) is two-dimensional. Therefore, in order to apply the results of Section 1.9, we reduce the problem to two independent problems: for even eigenoscillations [i.e., when $\varphi(-\phi)=\varphi(\phi)$] and for odd ones [$\varphi(-\phi)=-\varphi(\phi)$]. Thus, for odd φ, operator-valued function $\tilde{K}(\lambda,a)$ can be represented in the form

$$\tilde{K}(\lambda,l)\tilde{\varphi}=K_{NM}(\lambda,l)\tilde{\varphi}+\frac{\mu_{NM}}{\lambda-\lambda_{NM}}(\tilde{\varphi},\varphi_N)\varphi_N,$$

and for odd φ,

$$\tilde{K}(\lambda,l)\tilde{\varphi}=K_{NM}(\lambda,l)\tilde{\varphi}+\frac{\mu_{NM}}{\lambda-\lambda_{NM}}(\tilde{\varphi},\psi_N)\psi_N.$$

Applying formula (1.143), we obtain the CN in the vicinity of λ_{NM}. For even oscillations,

$$\lambda^*_{NM}=\frac{j'_{N,M}}{\sqrt{\varepsilon_2}R}+\beta\frac{1}{\pi j'_{N,M}\sqrt{\varepsilon_2}R}\frac{1}{1-\frac{N^2}{j'^2_{N,M}}}\frac{\varepsilon_2}{\varepsilon_1+\varepsilon_2}$$

$$-\beta^2\frac{1}{\pi j'_{N,M}\sqrt{\varepsilon_2}R}\frac{1}{1-\frac{N^2}{j'^2_{N,M}}}\frac{\varepsilon_2}{\varepsilon_1+\varepsilon_2}\left(\frac{\ln 2}{\pi}+\frac{\Re C_{NM}\left(\frac{j'_{N,M}}{\sqrt{\varepsilon_2}R}\right)}{\varepsilon_1+\varepsilon_2}\right)$$

$$-i\beta^2\frac{1}{\pi^3 j'^3_{N,M}\sqrt{\varepsilon_2}R}\frac{2}{1-\frac{N^2}{j'^2_{N,M}}}\frac{\varepsilon_2^2}{(\varepsilon_1+\varepsilon_2)^2}\sum_{n=0}^{\infty}\frac{\delta_n}{\left(J_n'\left(j'_{N,M}\sqrt{\frac{\varepsilon_1}{\varepsilon_2}}R\right)\right)^2+\left(N_n'\left(j'_{N,M}\sqrt{\frac{\varepsilon_1}{\varepsilon_2}}R\right)\right)^2}$$

$$+O(\beta^3),\quad\text{as}\quad\beta\to 0.$$

For odd oscillations, the CN in the vicinity of the same pole λ_{NM} can be obtained using formula (1.145). The operator-valued function $N(\lambda,l)$ at real λ can be represented as

$$N(\lambda,l)=\nu(\lambda,l)P_2,$$

where

$$\nu(\lambda,l)=\frac{\lambda^2R^2}{4\pi}(\varepsilon_1^2+\varepsilon_2^2)-i\frac{1}{\ln l}\frac{1}{\pi^2(\lambda R)^2}\sum_{n=1}^{\infty}\frac{n^2}{\left(J_n'(\lambda\sqrt{\varepsilon_1}R)\right)^2+\left(N_n'(\lambda\sqrt{\varepsilon_1}R)\right)^2}.$$

Then, as follows from formula (1.147), the CN is expressed by the formula

$$\lambda^*_{NM} = \frac{j'_{N,M}}{\sqrt{\varepsilon_2}R} + l^2 \frac{1}{2j'_{N,M}\sqrt{\varepsilon_2}R} \frac{N^2}{1 - \dfrac{N^2}{j'^2_{N,M}}} \frac{\varepsilon_2}{\varepsilon_1 + \varepsilon_2}$$

$$-il^4 \frac{N^2}{1 - \dfrac{N^2}{j'^2_{N,M}}} \frac{1}{2\sqrt{\varepsilon_2}R} \frac{\varepsilon_2^2}{(\varepsilon_1 + \varepsilon_2)^2} \frac{1}{\pi j'^3_{N,M}} \sum_{n=1}^{\infty} \frac{n^2}{\left(J'_n(j'_{N,M}\sqrt{\frac{\varepsilon_1}{\varepsilon_2}})\right)^2 + \left(N'_n(j'_{N,M}\sqrt{\frac{\varepsilon_1}{\varepsilon_2}})\right)^2} + \rho(l),$$

where $\rho(l)$ is a complex quantity such that

$$\Re\rho(l) = O(l^4 \ln l) \quad \text{and} \quad \Im\rho(l) = O(l^6) \quad \text{as} \quad l \to 0.$$

As expected, the CNs are complex with negative imaginary part.

4.5 The integral and summation equations for the strip problems

In the present section, we will apply the technique of summation operators to solving the strip diffraction problem *Hi* formulated in Section 4.1. We will reduce, following [25, 27], the problem in question to a canonical Fredholm summation equation, where the operator differs from that obtained in the previous sections and in [10, 20, 21] (see also the survey [7]) using the expansion in the Chebyshev polynomials. In fact, the logarithmic integral equation of the problem [in which integration is performed over an arbitrary (piecewise smooth unclosed finite) curve] is reduced to a summation equation in the form of an infinite system using the trigonometric basis rather than the Chebyshev system.

We will consider both homogeneous and inhomogeneous problems *H* and *Hi* (according to the notation of Section 4.1). In order to obtain the boundary integral equations with a logarithmic singularity of the kernel, we will replace in the latter case homogeneous Neumann boundary value condition (4.2) on Γ by the inhomogeneous Dirichlet condition

$$u(\mathbf{x}) = f(\mathbf{x}), \quad \mathbf{x} \in \Gamma, \tag{4.46}$$

Such a replacement is possible due to the reciprocity principle [22]. Here, f is a given continuous function defined on Γ that enters the formulation of problem *Hi* of the plane wave diffraction by a curved strip Γ given in Section 4.1.

Following [25], we will reduce problem *Hi* with condition (4.46) to coupled summation equations that admit an equivalent regularization and the reduction to a canonical Fredholm operator equation.

Every solution of inhomogeneous boundary value problem *Hi* admits the representation in the form of a (generalized) single-layer potential

$$u(\mathbf{x}) = \int_{\Gamma} \mathcal{E}(\mathbf{x} - \mathbf{y}) Z(\mathbf{y}) ds_{\mathbf{y}}, \tag{4.47}$$

where

$$\mathcal{E}(\mathbf{x} - \mathbf{y}) = \frac{i}{4} H_0^{(1)}(k|\mathbf{x} - \mathbf{y}|),$$

and $Z(\mathbf{y})$ is a certain function defined on Γ such that

$$Z(\mathbf{y}) = \sqrt{s(\mathbf{y},\mathbf{y}_1^e) \cdot s(\mathbf{y},\mathbf{y}_2^e)} H(\mathbf{y}), \quad \mathbf{y} \in \Gamma. \tag{4.48}$$

Here, $s(\mathbf{x},\mathbf{y})$ is the length of the arc measured from $\mathbf{x}$ to $\mathbf{y}$, and $H(\mathbf{y})$ belongs to the Hölder class with respect to the $s(\mathbf{x},\mathbf{y})$-norm. If Γ is a closed contour, condition (4.48) is ignored, and $Z(\mathbf{y})$ belongs to the Hölder class on Γ.

It is shown in [31] that potential (4.47) satisfies the Reichardt condition (4.11) (and the Sommerfeld radiation condition for real $k > 0$) when $k \neq 0$ is an arbitrary complex number.

Applying boundary condition (4.46) and the technique of Chapter 2, one can show that (unknown) density $Z(\mathbf{y})$ satisfies the integral equation

$$\int_\Gamma \mathcal{E}(\mathbf{x}-\mathbf{y}) Z(\mathbf{y})\, ds_{\mathbf{y}} = f(\mathbf{x}), \quad \mathbf{x} \in \Gamma. \tag{4.49}$$

Assume that a simple-connected contour Γ may be complemented to a closed contour Γ' without points of self-intersection. Then, Γ' is a topological equivalent of the unit circle

$$Q[-\pi,\pi] = \{\exp(j\theta),\ \theta \in [-\pi,\pi]\},$$

and there exist continuous functions $x = x(\theta)$ and $y = y(\theta)$ determining the one-to-one correspondence between Q and Γ': $[x(\theta), y(\theta)] \in \Gamma'$, $\theta \in [-\pi,\pi]$. Assume that S' is sufficiently smooth, namely, $x(\theta)$ and $y(\theta)$ admit a 2π-periodical continuations on the whole axis, which are once continuously differentiable, and their second and third derivatives are piecewise continuous functions.

Let the prototype of Γ correspond to the closed interval $\theta \in [-d,d]$ for a certain $d \in [0,\pi]$, and let θ_1 and θ_2 be the prototypes of points $\mathbf{x},\mathbf{y} \in S$. Then, (4.49) may be rewritten as

$$\int_{-d}^{d} \mathcal{E}[R(\theta_1,\theta_2)] Z(\theta_2) d\theta_2 = G(\theta_1), \quad \theta_1 \in (-d,d), \tag{4.50}$$

where $G(\theta_1)$ is the known function and $R(\theta_1,\theta_2)$ is the distance between images θ_1 and θ_2 on the (x,y)-plane.

Define the function

$$Z^0(\theta) = \begin{cases} Z(\theta), & \theta \in (-d,d), \\ 0, & \theta \in [-\pi,\pi] \setminus (-d,d). \end{cases} \tag{4.51}$$

It follows from (4.50) and (4.51) that $Z^0(\theta)$ satisfies the equation

$$\int_{-\pi}^{\pi} \mathcal{E}[R(\theta_1,\theta_2)] Z^0(\theta_2) d\theta_2 = G(\theta_1), \quad \theta_1 \in (-d,d). \tag{4.52}$$

Here,

$$\mathcal{E}[R(\theta_1,\theta_2)] = i\frac{2}{\pi}[\ln|2\sin\Theta_{12}| + H(\theta_1,\theta_2)], \quad \Theta_{12} = \frac{\theta_1-\theta_2}{2}. \tag{4.53}$$

Below, we will apply the assumption that the Fourier coefficients h_{mn} of the double Fourier series

$$H(\theta,\tau) = \sum_{m=-\infty}^{\infty} \sum_{n=-\infty}^{\infty} h_{mn} \exp[j(m\theta + n\tau)], \quad \theta,\tau \in [-\pi,\pi], \tag{4.54}$$

are known, to reduce (4.52) to a system of summation equations by the method developed in [25, 27]. This method is different from the one described in Sections 4.1 and 4.2, which is based on the Fourier–Chebyshev expansions in terms of the Chebyshev polynomials. Here, we will use, in place of formula (1.51), another similar property of the logarithmic kernel:

$$\ln|2\sin\Theta_{12}| = -\frac{1}{2}\sum_{n=-\infty,n\neq 0}^{\infty}\frac{1}{|n|}\exp(jn\theta_1)\exp(-jn\theta_2), \quad \theta_1,\theta_2\in[-\pi,\pi]. \tag{4.55}$$

We will look for the solution $Z^0(\tau)$ in the form of the Fourier series

$$Z^0(\tau) = \sum_{n=-\infty}^{\infty} z_n\exp(jn\tau), \quad \tau\in[-\pi,\pi] \tag{4.56}$$

[similar to (1.62)]. Substituting (4.53)–(4.56) to equation (4.52) and taking into account the orthogonality and completeness of $\{\exp(nj\tau)\}_{n=-\infty}^{\infty}$, we obtain, repeating procedure (1.62)–(1.64) of the reduction of an integral equation (1.54) to (3.8), an infinite system

$$\begin{aligned}\sum_{n=-\infty,n\neq 0}^{\infty}\frac{1}{|n|}z_n\exp(jn\theta) - 2\sum_{n=-\infty}^{\infty} z_n\sum_{p=-\infty}^{\infty} h_{p,-n}\exp(jp\theta)\\ = \sum_{n=-\infty}^{\infty} g_n\exp(jn\theta), \quad \theta\in[-d,d],\end{aligned} \tag{4.57}$$

where $\{g_n\}_{n=-\infty}^{\infty}$ are the known Fourier coefficients of function $\frac{j}{2}G(\theta)$ smoothly continued on $[-\pi,\pi]$, satisfying the condition

$$\sum_{n=-\infty}^{\infty}(|n|+1)|g_n|^2 < \infty.$$

Equations (4.55) and (4.51) yield

$$\sum_{n=-\infty}^{\infty} z_n\exp(jn\tau) = 0, \quad \tau\in[-\pi,\pi]\setminus(-d,d). \tag{4.58}$$

(4.57) and (4.58) are the coupled summation equations with respect to the unknown infinite vector $\{z_n\}_{n=-\infty}^{\infty}$ of the Fourier coefficients satisfying the conditions

$$\sum_{n=-\infty}^{\infty}(|n|+1)^{-1}|z_n|^2 < \infty. \tag{4.59}$$

Note that (4.59) differs from the definition of the spaces h_p given in Section 3.1. But we will not directly consider (4.57) and (4.58) in the spaces of infinite sequences determined by the corresponding condition (4.59). Following [25, 27], we will regularize this system of coupled summation equations to obtain a canonical second-kind summation system in the space

$$h_0 = l_2 = \{\{z_n\}_{n=0}^{\infty} : \sum_{n=0}^{\infty}|z_n|^2 < \infty\}.$$

The following theorem proved in [25] gives the rate of decrease for coefficients h_{mn} and, hence, guarantees the applicability of the regularization procedure.

Theorem 41 ([25]) *Assume that functions $x(\theta)$ and $y(\theta)$ that determine the parametrization of the integration contour admit 2π-periodical continuations on the whole axis, which are once continuously differentiable, and their second and third derivatives are piecewise continuous functions. Then,*

$$\sum_{n=-\infty}^{\infty} \sum_{m=-\infty}^{\infty} (|n|+1)^2(|m|+1)^2|h_{nm}|^2 < \infty. \tag{4.60}$$

Generalization of the approach developed in [27] allows us to regularize system (4.57) and (4.58) by the technique described in [25, 26] (which is different from the method of the Riemann-Hilbert problem [27]), and to obtain the equivalent infinite algebraic system of the second kind

$$(I + D + BH)\mathbf{z} = B\mathbf{g}, \tag{4.61}$$

where I is the identity operator. The calculations of the matrix elements of summation operators D, B, and H are performed in [25, 26]. Here, we will limit ourselves to presenting the final relationships:

$$D = \begin{pmatrix} D^{(+)} & 0 \\ 0 & 0 \end{pmatrix}, \quad B = \begin{pmatrix} B^{(+)} & 0 \\ 0 & B^{(-)} \end{pmatrix}, \quad H = \begin{pmatrix} H^{(+,+)} & H^{(+,-)} \\ H^{(-,+)} & H^{(-,-)} \end{pmatrix}$$

are the summation operators such that

$$D^{(+)} = \{d_{sn}^{(+)}\}_{s,n=0}^{\infty}, \quad B^{(+)} = \{b_{sn}^{(\pm)}\}_{s,n=0}^{\infty}, \quad H^{(\pm,\pm)} = \{h_{sn}^{(\pm,\pm)}\}_{s,n=0}^{\infty},$$

where

$$d_{00}^{(+)} = -1 - \sqrt{2}\ln\frac{1-t_0}{2}, \quad d_{s0}^{(+)} = -\sqrt{2}p_s \ (s \geq 1), \quad d_{sn}^{(+)} = 0 \ (n \geq 1),$$

$$b_{00}^{(+)} = 0, \quad b_{0n}^{(+)} = p_n \ (n \geq 1), \quad b_{s0}^{(+)} = 0 \ (s \geq 1),$$

$$b_{sn}^{(+)} = \delta_{sn} - Q_{s-1,n-1}^{(1,0)}(t_0), \quad b_{sn}^{(-)} = \delta_{sn} - Q_{s-1,n-1}^{(0,1)}(t_0) \ (s, n \geq 1),$$

$$h_{sn}^{(+,\pm)} = -\frac{\tau_s\tau_n}{1+\delta_{s0}}[(h_{s,-n} \pm h_{sn}) + (h_{-s,-n} \pm h_{-s,n})] \ (s \geq 0, \ n \geq 1),$$

$$h_{sn}^{(-,\pm)} = -\tau_s\tau_n[(h_{s,-n} \pm h_{sn}) - (h_{-s,-n} \pm h_{-s,n})] \ (s \geq 1, \ n \geq 1).$$

Here,

$$p_n = \frac{1}{n}(1+t_0)\hat{P}_{n-1}^{(0,1)}(t_0), \quad t_0 = \cos d, \quad \tau_n = \sqrt{n} \ (n \geq 1),$$

and $\hat{P}_n^{(\alpha,\beta)}(x)$ are the Jacobi polynomials (see [25]). The vectors of the right-hand sides and the unknowns in (4.61) have the form

$$\mathbf{g} = (\mathbf{g}^{(+)}, \mathbf{g}^{(-)})^T, \quad \mathbf{z} = (\mathbf{z}^{(+)}, \mathbf{z}^{(-)})^T,$$

$$\mathbf{g}^{(+)} = \{g_n^{(+)}\}_{n=0}^{\infty}, \quad \mathbf{g}^{(-)} = \{g_n^{(-)}\}_{n=1}^{\infty}\mathbf{z}^{(+)} = \{z_n^{(+)}\}_{n=0}^{\infty}, \quad \mathbf{z}^{(-)} = \{z_n^{(-)}\}_{n=1}^{\infty},$$

$$z_n^{(+)} = \frac{z_n + z_{-n}}{\tau_n(1+\delta_{n0})}, \quad z_n^{(-)} = \frac{z_n - z_{-n}}{\tau_n} g_n^{(+)} = \frac{\tau_n(g_n + g_{-n})}{1+\delta_{n0}}, \quad g_n^{(-)} = \tau_n(g_n - g_{-n}).$$

Let us introduce the space of vector-columns $l_2^{(0)} = l_2 \bigotimes l_2$ and formulate, referring to the proof given in [25], the following statement.

Theorem 42 ([25]) *Assume that conditions (4.60) hold. Then, (4.61) is the second-kind algebraic system in $l_2^{(0)}$, i.e., it is reduced to the form*

$$(I + K)\mathbf{z} = \mathbf{b},$$

where $\mathbf{z}, \mathbf{b} \in l_2^{(0)}$, and K is a compact operator in $l_2^{(0)}$.

We have thus carried out the reqularization of integral equation (4.52) by reducing it to a canonical Fredholm equation with infinite-matrix operator $I + K$. This procedure may be literally repeated (for every complex $k \neq 0$) for the corresponding homogeneous equation to obtain the operator equation

$$[I + D + BH(k)]\mathbf{z} = \mathbf{o}, \quad \mathbf{z} \in l_2^{(0)}, \tag{4.62}$$

similar to (4.61) [with $\mathbf{g} = \mathbf{o}$, where $\mathbf{o}$ is a zero element of $l_2^{(0)}$], where D is a compact operator in $l_2^{(0)}$, B is a bounded operator in $l_2^{(0)}$, and the properties of operator-valued function $H(k)$ are stated below.

Lemma 16 ([25, 26]) *The compact operator-valued function $H(k) : l_2^{(0)} \to l_2^{(0)}$ is defined in Λ, where Λ is the Riemann surface of the analytical continuation of the fundamental solution $\mathcal{E} = -\frac{i}{4}H_0^{(1)}(k|\mathbf{x} - \mathbf{y}|)$ (or, what is the same, of function $\ln k$) of the two-dimensional Helmholtz equation $\mathcal{H}(k)u = 0$ with respect to spectral parameter k.*

Lemma 17 ([25, 26]) *$H(k)$ is a holomorphic operator-valued function in Λ.*

Following [25, 26], we will denote by F the linear transformation (described in more detail in [25]), which establishes a correspondence between a solution u of boundary eigenvalue problem H and a (nontrivial) solution $\mathbf{z}$ of (4.62). Then, the following equivalence theorem is valid.

Theorem 43 ([25, 26]) *Every solution u of boundary-eigenvalue problem H [with homogeneous condition (4.46)] defines a solution $\mathbf{z} = F(u) \in l_2^{(0)}$ of equation (4.62). There exists the inverse transformation such that for every solution $\mathbf{z}$ of (4.62), the function $u = F^{-1}\mathbf{z}$ is a solution of problem H.*

Equation (4.62) is a problem on CNs for a canonical Fredholm holomorphic operator-valued function $I + D + BH(k)$. Due to the results [51], boundary eigenvalue problem H has only a trivial solution for all $k \in \{k : 0 < \arg k < \pi\}$. Hence, for such k, there exists a bounded operator $[I + D + BH(k)]^{-1}$ which, in accordance with theorem 25, is a meromorphic operator-valued function in Λ with not more than a countable set of poles. These statements together with the results of theorem 43 enable us to prove the following results.

Theorem 44 ([25, 26]) *Fundamental frequencies and corresponding eigenvalues $\lambda \in \sigma_H(\Delta)$ of boundary eigenvalue problem H form not more than a countable set of (isolated) points in Λ.*

Theorem 45 ([25, 26]) *Inhomogeneous boundary value problem H with condition (4.46) [as well as integral equation (4.52) with a logarithmic singularity of the kernel] is uniquely solvable for all $\lambda \notin \sigma_H(\Delta)$.*

4.6 Summation equations in the problem on eigenfrequencies of a rectangular slotted resonator

Using the method set forth in Chapter 2 and Section 4.3.3, we reduce the problem on eigenfrequencies in a rectangular slotted resonator with an asymmetrically located slot $S = \{(x,y): d-w < x < d+w,\ y=0\}$ to the homogeneous integral equation

$$K(\lambda)\varphi \equiv \int_{d-w}^{d+w} K(\lambda;x,x_0)\varphi(x_0)dx_0 = 0, \quad x_0 \in S = (d-w, d+w), \tag{4.63}$$

where

$$K(\lambda;x,x_0) = \varepsilon_1 G_1(\lambda;x,0,x_0,0) + \varepsilon_2 G_2(\lambda;x,0,x_0,0). \tag{4.64}$$

According to definitions of Section 1.6, we can consider (4.63) as a problem on CNs for the Fredholm integral operator-valued function of complex variable λ.

Introducing the new variables $s = (x_0-d)/w$ and $t=(x-d)/w$, we pass to the standard interval of integration $(-1,1)$ in (4.63) to obtain [setting $\varphi(x_0)=\psi(s)$]

$$K(\lambda,\vec{a})\psi \equiv \int_{-1}^{1} \tilde{K}(\lambda,\vec{a};s,t)\frac{\psi(s)}{\sqrt{1-s^2}}ds = 0, \tag{4.65}$$

where

$$\tilde{K}(\lambda,\vec{a};s,t) = \frac{\varepsilon_1+\varepsilon_2}{\pi}\ln\frac{1}{|s-t|} + \frac{\varepsilon_1+\varepsilon_2}{\pi}\ln\frac{a}{\pi w}$$

$$-\sum_{j=1}^{2}\left\{\frac{\varepsilon_j}{\pi}\left[\ln 2\frac{\sin\frac{w}{2a}|t-s|}{\pi w|t-s|} + \ln 2\sin\frac{\pi}{2a}(w(t+s)+2d)\right]\right\}$$

$$+\beta_0(\lambda) + \sum_{n=1}^{\infty}\left(\beta_n(\lambda) - \frac{(\varepsilon_1+\varepsilon_2)a}{\pi n}\right)\cos\frac{\pi n}{a}(wt+d)\cos\frac{\pi n}{a}(ws+d)$$

$$= \frac{\varepsilon_1+\varepsilon_2}{\pi}\ln\frac{1}{|s-t|} + \frac{\varepsilon_1+\varepsilon_2}{\pi}\ln\frac{1}{w} + M(\lambda,\vec{a};s,t), \quad s\in[-1,1],$$

$$\beta_n(\lambda) = \frac{2}{a(1+\delta_{0n})}\sum_{j=1}^{2}\varepsilon_j\frac{\coth(b_j\sqrt{(\pi n a^{-1})^2-\lambda\varepsilon_j})}{\sqrt{(\pi n a^{-1})^2-\lambda\varepsilon_j)}}.$$

Here, $\vec{a} = (w, a_1, a_2, b_1, b_2, \varepsilon_1, \varepsilon_2)$ is the vector of nonspectral parameters of the problem and $K(\lambda,\vec{a}) : L_2^{(1)}(-1,1) \to \widetilde{W}_2^1(-1,1)$ is the Fredholm finite-meromorphic integral operator-valued function. Note that we have separated explicitly the weight function $1/\sqrt{1-x^2}$.

Rewrite homogeneous integral equation (4.65) in the operator form

$$K(\lambda)\psi \equiv c_0 L\psi + N(\lambda)\psi = 0,$$

where

$$L\psi(s) = -\frac{1}{\pi}\int_{-1}^{1}\ln|s-t|\frac{\psi(s)}{\sqrt{1-s^2}}ds,$$

$$\Phi_N(t) = N(\lambda)\psi = \frac{1}{\pi}\int_{-1}^{1} N(\lambda; s, t)\frac{\psi(s)}{\sqrt{1-s^2}}ds.$$

According to the results of Section 3.4, the set of functions Φ that can be expanded in the Fourier–Chebyshev series with the vector of coefficients

$$\xi = (\xi_0, \xi_1, \dots \xi_n, \dots) \in h_2 = \{\xi : \sum_{n=0}^{\infty} |\xi_n|^2 n^2 < \infty\}$$

is isomorphic to $\widetilde{W}_2^1$. Expand function $\psi(s)$ in the Fourier–Chebyshev series

$$\psi(x) = \frac{1}{\sqrt{2}}T_0(x)\xi_0 + \sum_{n=1}^{\infty} \xi_n T_n(x), \quad -1 \le x \le 1, \tag{4.66}$$

where $\xi \in h_2$. $\Phi_N(t)$ is a continuous function; therefore, it can also be expanded in the Fourier–Chebyshev series

$$\Phi_N(t) = \frac{1}{2}T_0(t)\eta_0 + \sum_{n=1}^{\infty} \eta_n T_n(t), \quad -1 \le t \le 1, \tag{4.67}$$

$$\eta_n = \frac{2}{\pi^2}\int_{-1}^{1}\int_{-1}^{1} N(\lambda; s, t)\psi(s)T_n(t)\frac{ds}{\sqrt{1-s^2}}\frac{dt}{\sqrt{1-t^2}}. \tag{4.68}$$

Substituting (4.66) into (4.68) and using the properties of the Chebyshev polynomials we obtain the equation, which is equivalent to (4.65) in $\widetilde{W}_2^1$:

$$c_0 \ln 2 \frac{\xi_0}{\sqrt{2}}T_0(t) + c_0 \sum_{n=1}^{\infty} n^{-1}\xi_n T_n(t) + \frac{\eta_0}{2}T_0(t) + \sum_{n=1}^{\infty} \eta_n T_n(t) = 0, \quad |t| \le 1.$$

In its turn, the latter equation is equivalent to the infinite system

$$\begin{cases} c_0\xi_0 \frac{1}{\sqrt{2}} \ln 2 + \frac{\eta_0}{2} = 0, \\ c_0 n^{-1}\xi_n + \eta_n = 0, \ n \ge 1, \end{cases}$$

which yields the summation-type equation

$$K(\lambda)\xi = c_0 L\xi + A(\lambda)\xi = 0, \quad \xi = (\xi_0, \xi_1, \dots, \xi_n, \dots) \in h_2, \tag{4.69}$$

where

$$L = \begin{pmatrix} \ln 2 & & & & & \\ & 1 & & & 0 & \\ & & \frac{1}{2} & & & \\ & & & \ddots & & \\ & 0 & & & n^{-1} & \\ & & & & & \ddots \end{pmatrix}, \quad A = \|a_{nj}(\lambda)\|_{n,j=0}^{\infty};$$

$$a_{nj} = \varepsilon_{nj} \iint_{\Pi_1} \frac{1}{\pi^2} N(\lambda; s, t)T_n(t)T_j(s)\frac{dx}{\sqrt{1-s^2}}\frac{dt}{\sqrt{1-t^2}};$$

$$\varepsilon_{nj} = \begin{cases} 1, & n = j = 0 \\ 2, & n \le 1,\ j \le 1, \\ \sqrt{2}, & \text{for all other } n,\ j \end{cases}$$

We look for the CNs $\lambda^* = \lambda^*(\vec{a})$ as zeros of the infinite determinant of the infinite matrix of this summation-type equation. According to [6], the infinite determinant exists for the Koch matrix

$$E + C(\lambda) = \{\delta_{nj} + c_{nj}(\lambda)\}_{n,j=0}^{\infty}, \tag{4.70}$$

where coefficients c_{nj} satisfy the conditions

$$\sum_{n=0}^{\infty} |c_{nn}| < \infty, \quad \sum_{n,j=0}^{\infty} |c_{nj}|^2 < \infty. \tag{4.71}$$

In order to reduce matrix (4.69) to the form (4.70), we multiply it on the left by matrix $c_0^{-1}L^{-1}$ to obtain

$$\tilde{K} = c_0^{-1}KL^{-1} = E + c_0^{-1}AL^{-1} = E + C, \tag{4.72}$$

where

$$C = \{c_{nj}\}_{n,j=0}^{\infty}, \quad c_{nj} = \begin{cases} \dfrac{a_{n0}}{c_0 \ln 2}, \\ \\ \dfrac{j a_{nj}}{c_0}, & j \ne 0. \end{cases} = \begin{cases} \dfrac{a_{n0}}{(\varepsilon_1 + \varepsilon_2)\ln 2}, \\ \\ \dfrac{j a_{nj}}{\varepsilon_1 + \varepsilon_2}, & j \ne 0. \end{cases}$$

Coefficients a_{nj} have the form

$$a_{nj} = \frac{\varepsilon_{nj}}{\pi^2}\Bigg(\sum_{i=1}^{2} \int_{-1}^{1}\int_{-1}^{1} \varepsilon_i \Big(\ln\frac{a}{\pi w} - \ln 2a \frac{\sin\frac{\pi w}{2a}|s-t|}{\pi w|s-t|}$$

$$- \ln 2\sin\big(\frac{\pi}{2a}(w(s+t)+2d)\Big) T_n(t)T_j(s)\frac{dsdt}{\sqrt{1-s^2}\sqrt{1-t^2}}$$

$$+\pi \int_{-1}^{1}\int_{-1}^{1}\Big(\beta_0(\lambda) + \sum_{k=1}^{\infty}\Big(\beta_k(\lambda) - \frac{(\varepsilon_1+\varepsilon_2)a}{\pi k}\Big)\cos\frac{\pi k}{a}(wt+d)\cos\frac{\pi k}{a}(ws+d)\Big)$$

$$\times T_n(t)T_j(s)\frac{dsdt}{\sqrt{1-s^2}\sqrt{1-t^2}}\Bigg).$$

The last integral is calculated explicitly, so that the final formulas are

$$c_{nj} = \gamma_{nj}\frac{\varepsilon_{nj}}{\pi^2}\sum_{i=1}^{2}\Bigg\{\int_{-1}^{1}\int_{-1}^{1} \varepsilon_i\Big(\ln\frac{a}{\pi w} - \ln 2a\frac{\sin\frac{\pi w}{2a}|s-t|}{\pi w|s-t|}$$

$$- \ln 2\sin\big(\frac{\pi}{2a}(w(s+t)+2d)\big)\Big)T_n(t)T_j(s)\frac{dsdt}{\sqrt{1-s^2}\sqrt{1-t^2}}$$

$$+\pi^3\beta_0(\lambda)\delta_{n0}\delta_{j0} + \pi^3\sum_{k=1}^{\infty}\Big(\beta_k(\lambda) - \frac{(\varepsilon_1+\varepsilon_2)a}{\pi k}\Big)J_n(\pi k\frac{w}{a})J_j(\pi k\frac{w}{a})C_{nk}C_{jk}\Bigg),$$

where $J_k(x)$ are the Bessel functions;

$$C_{jk} = \begin{cases} (-1)^{\frac{j}{2}}\cos(\pi k\frac{d}{a}), & j \text{ even}, \\ \\ (-1)^{\frac{j+1}{2}}\sin(\pi k\frac{d}{a}), & j \text{ odd}, \end{cases}$$

$$\gamma_{nj} = \begin{cases} \dfrac{1}{(\varepsilon_1 + \varepsilon_2)\ln 2}, \\ \\ \dfrac{j}{\varepsilon_1 + \varepsilon_2}, \quad j \neq 0. \end{cases}$$

One can verify directly that conditions (4.71) are valid for coefficients c_{nj}; consequently, there exists the (infinite) determinant of (4.72).

The CNs $\lambda^* = \lambda^*(\vec{a})$ (eigenvalues of problem H) are zeros of the determinant of matrix (4.73):

$$F(\lambda^*, \vec{a}) \equiv \det[E + C(\lambda^*, \vec{a})] = 0. \tag{4.73}$$

For every fixed $\vec{a}$ such that $S \subset (0, a)$, $F(\lambda^*, \vec{a})$ is a meromorohic function with poles β^i_{nk}. From Theorem 18, it follows that in the interval between every two neighboring poles β^i_{nk}, function $F(\lambda, \vec{a})$ has at least one real zero (CN).

Using the terminology of Section 1.6, we can separate a simple pole pencil of $K(\lambda)$ in the vicinity of any pole $\lambda^{(i)}_{NM}$:

$$K(\lambda)\xi = c_0 L\xi + \frac{A_{-1}(\lambda)\xi}{\lambda - \lambda^{(i)}_{NM}} + N(\lambda)\xi = 0;$$

the elements of matrix $A_{-1}(\lambda)$ have the form

$$a_{nj}(\lambda) = \pi\varepsilon_{nj}\tilde{\beta}^{(i)}_{NM}(\lambda)C_{nN}C_{jN}J_n(\pi N\frac{w}{a})J_j(\pi N\frac{w}{a}),$$

$$\tilde{\beta}^{(i)}_{NM}(\lambda) = -\frac{2 - \delta_{0N}}{ab_i} \frac{\cos\sqrt{z}(\sqrt{z} + \pi M)}{\sqrt{z}} \frac{\sqrt{z} - \pi M}{\sin\sqrt{z}},$$

where

$$z = -\varepsilon_i b_i^2 \Big(\frac{\pi^2 N^2}{\varepsilon_i a^2} - \lambda\Big), \quad i = 1, 2.$$

Thus, we have applied the basis of Chebyshev polynomials to reduce a canonical problem concerning the determination of eigenfrequencies of a rectangular slotted resonator to logarithmic integral equation (4.63), summation equation (4.64), and, finally, to a functional equation (4.73). Actual calculations are performed for a truncated matrix of a finite order. This approach can be extended to slotted resonators formed by domains of arbitrary shape, if one separates a logarithmic singularity in kernel (4.64) and expands regular terms of the traces of Green's functions in a double trigonometric Fourier or Chebyshev series.

Bibliography

1. N.I. Muskhelishvili. *Singular integral equations.* Nauka, Moscow, (1962) (in Russian)

2. N.P. Vekua. *Systems of singular integral equations*, Nauka, Moscow, (1970) (in Russian).

3. F.D. Gakhov, *Boundary-value problems.* Nauka, Moscow, (1977) (in Russian).

4. S.G. Samko. *Hypersingular integrals and their applications.* Rostov-na-Donu Univ. Press, Rostov-na-Donu, (1984) (in Russian).

5. B.R. Vainberg. *Asymptotic methods in the equations of mathematical physics.* Moscow Univ. Press, Moscow (1977) (in Russian).

6. I. Gohberg, N. Krupnik. *One-dimensional linear singular integral equations.* Birkhäser, Basel, (1992).

7. M.S. Krasnosel'skii *et al. Integral operators in the spaces of integrable functions.* Nauka, Moscow, (1969) (in Russian).

8. *Numerical solution to integral equations.* Ed. by M. Golberg. Plenum Press, New York, (1990)

9. I.I. Vorovich *et al.*, *Nonclassical mixed problems of the elasticity theory.* Nauka, Moscow, (1974) (in Russian).

10. Z.T. Nazarchuk. *Numerical treatment of the wave diffraction by cylindrical structures.* Naukova Dumka, Kiev, (1989) (in Russian).

11. M. Costabel. Boundary integral operators of Lipschitz domains: elementary results. *SIAM J. Math. Anal.*, **19**, 613–626 (1988).

12. G.C. Hsiao, W. Wendland. A finite element method for some integral equations of the first kind. *J. Math. An. Appl.*, **58**, 449–481 (1977).

13. M. Costabel, E. Stephan. A direct boundary integral equation method for transmission problems. *J. Math. An. Appl.*, **106**, 367–413 (1985).

14. W. Wendland, E. Stephan, G.C. Hsiao. On the integral equation method for the plane mixed boundary-value problem of the Laplacian, *Math. Meth. Appl. Sc.*, **1**, 265–321 (1979).

15. V. van Splunter, P.M. van der Berg. Spectral theory of diffraction of electromagnetic waves by a strip in the plane interface of two semi-infinite media. *Canad. J. Phys.*, **57**, 1148–1156 (1979).

16. T. von Petersdorff. Boundary integral equations for mixed Dirichlet, Neumann, and transmission problems. *Math. Meth. Appl. Sc.*, **11**, 185–213 (1989).

17. E. Stephan. Boundary integral equations for mixed boundary value problems. *Math. Nachr.*, **131**, 167–199 (1987).

18. E. Stephan. Boundary integral equations for screen problems in R^3. *J. Integr. Eq. Oper. Th.*, **10**, 236–257 (1987).

19. M. Durand. Layer potentials and boundary-value problems for the Helmholtz equation in the complement of a thin obstacle. *Math. Meth. Appl. Sc.*, **5**, 389–421 (1983).

20. P. Wolfe. On the inverse of an integral operator. *Proc. Am. Math. Soc.*, **25**, 443–448 (1970).

21. R.A. Hurd, Y. Hayashi. Low-frequency scattering by a slit in a conducting plane. *Radio. Sc.*, **15**, 1171–1178, (1980).

22. H. Hönl, A.W. Maue, K. Westpfahl. *Theorie der Beugung*, Springer-Verlag, Berlin, (1961).

23. V.P. Shestopalov. *Method of Riemann-Hilbert problem in the theory of diffraction and propagation of electromagnetic waves*, Khar'kov Univers. Press, Khar'kov, (1971) (in Russian).

24. V.P. Shestopalov. *Spectral theory and excitation of open structures.* Naukova Dumka, Kiev, (1987) (in Russian).

25. A.E. Poyedinchuk, Yu.A. Tuchkin, V.P. Shestopalov. Diffraction by curved strips. *Trans. Inst. Elect. Eng. of Japan*, **113-A**, 139–196, (1993).

26. V.P. Shestopalov, Yu.V. Shestopalov. *Spectral theory and excitation of open structures*, Peter Peregrinus Ltd, London, (1996).

27. V.P. Shestopalov. *Summation equations in the modern theory of diffraction*, Naukova Dumka, Kiev, (1983) (in Russian).

28. V.G. Sologub. On a solution of a convolution-type integral equation with the finite limits of integration. *Zh. Vych. Mat.*, **41**, 837–854, (1971).

29. Yu.V. Shestopalov. Justification of the spectral method for calculating the normal waves of transmission lines. *Differents Uravn.*, **16**, 1504-1512 (1980).

30. A.E. Poyedinchuk. On a spectral theory of open two-dimensional resonators with dielectric inclusions. *Doklady AN USSR. Ser. A*, **8**, 46-48 (1983).

31. A.S. Ilyinsky, Yu.V. Shestopalov. *Applications of the methods of spectral theory in the problems of wave propagation*, Moscow Univ. Press, Moscow (1989) (in Russian).

32. A.S. Ilyinsky, Yu.G., Smirnov. Mathematical modeling of the propagation of electromagnetic waves in a slotted transmission line. *Zhurn. Vychiscl. Matem. i Matem. Fiz.*, **27**, 163–170, (1987).

33. A.S. Ilyinsky, G.Ya. Slepyan, A.Ya. Slepyan. *Propagation, scattering and dissipation of electromagnetic waves*, Peter Peregrinus Ltd., London, (1993).

34. A.S. Ilyinsky, Yu.G., Smirnov. *Electromagnetic wave diffraction by conducting screens*, VSP, Utrecht, (1998).

35. I.T. Gohberg, E.I. Sigal. Operator generalization of the logarithmic residue theorem and Rouchet's theorem. *Matematich Sbornik*, **84**, 607–629 (1971) (in Russian).

36. E. Sanchez-Palencia. *Nonhomogeneous media and vibration theory.* Springer-Verlag, Berlin, (1980).

37. E.V. Chernokozhin, Yu.V. Shestopalov. Fredholm property of an integral operator with the kernel having a weak singularity. *Vestnik MGU, Ser. 15*, **1**, 23–28 (1982).

38. Yu.V. Shestopalov. Application of the method of generalized potentials in some problems of wave propagation and diffraction. *Zhurn Vychisl. Matem. i Matem. Fiziki*, **30**, 1081–1092 (1990).

39. Yu.V. Shestopalov. On the theory of cylindrical resonators. *Math. Methods in the Appl. Sciences*, **14**, 335–375, (1991).

40. E.V. Chernokozhin. A method for calculating the characteristic numbers of an operator-valued function of a special kind. *Vestnik Mosk. Gos. Univ. Ser. 15.*, no. 3, 76–77 (1985).

41. V.A. Ilyin. Kernels of fractional order. *Matematich. Sbornik*, **69**, 319–351, (1956).

42. E.V. Chernokozhin, Yu.V. Shestopalov. On the solvability of boundary-value problems for the Helmholtz equation in an unbounded domain with a noncompact boundary. *Differents. Uravn.*, **34**, 546–553 (1998).

43. V.A. Sadovnichii. *Theory of operators.* Moscow Univ. Press, Moscow, (1989) (in Russian).

44. Yu.V. Shestopalov. *Method of normal waves in the theory of oscillations*, Moscow Univ. Press, Moscow, (1985) (in Russian).

45. E.V. Chernokozhin, Yu.V. Shestopalov. Mathematical methods for the study of wave scattering by open cylindrical structures. *J. Comm. Tech. Elec.*, **42**, 1211–1223 (1997).

46. Ya.N. Fel'd. Diffraction of electromagnetic waves by unclosed screens. *Radiotekh. Elektron.*, **20**, 28–38, (1975).

47. H. Reichardt. Ausstralungsbedingungen für die Wellengleichung. *An. Math. Semin. Univ. Hamburg*, **24**, 41–53, (1960).

48. V.D. Kupradze. On approximate solution to the problems of mathematical physics. *Uspekhi Matematicheskikh Nauk.* **22**, 59–134, (1967).

49. E.V. Chernokozhin, Yu.V. Shestopalov. Scattering frequencies of open slotted resonators. *Proc. 1995 Int. Symp. URSI Elect. Theory, St Petersburg, May 23-26, 1995* **1**, 73–75, (1995).

50. M. Reed, B. Simon. *Methods of modern mathematical physics. Vol. 1*, Academic Press, New York, (1972).

51. V.V. Lozhechko, Yu.V. Shestopalov. On problems of excitation of open cylindrical resonators. *Zhurn. Vychisl. Matemat. i Matemat. Fiz.*, **35**, 71–83, (1995).

52. P.K. Suetin. *Classical orthogonal polynomials.* Nauka, Moscow, (1979) (in Russian).

53. S. Paszkowski. *Zastosowania numeryczne wielomianów i szeregów Czebyszewa.* Państwowe Wydawnictwo Naukowe, Warszawa, (1975).

54. Y. Luke. *The special functions and their approximations.* Springer-Verlag, New York, (1969).

55. I.S. Gradshtein, I.M. Ryzhik. *Tables of integrals, sums, series and products.* Fizmatgiz, Moscow, (1965) (in Russian).

56. A.S. Ilyinsky, Yu.G. Smirnov. Analysis of mathematical models of microstrip transmission lines. *Computational Mathematics and Modelling.* **1**, 218-234 (1990).

57. A.S. Ilyinsky, Yu.G. Smirnov. Computational modelling of slot lines. *Problems of Applied Mathematics*, Moscow Univ. Press, Moscow, 127-138 (1989).

58. A.S. Ilyinsky, Yu.G. Smirnov. Numerical solution of the slot lines formed by rectangular waveguides with various cross- section. *Radiotekhnika i Electronika.* **34**, 908-916 (1989).

59. Yu.G. Smirnov. Method of singular integral operator-functions for the transmission line problem. *Electromagnetics.* **14**, 145-156 (1993).

60. R. Kress. *Linear integral equations.* Applied Mathematical Science. **82**, Springer-Verlag, New York, (1989).

61. A.S. Ilyinsky, V.V. Kravtsov, A.G. Sveshnikov. *Mathematical models in electrodynamics.* Vissh. Shkola, Moscow, (1991) (in Russian).

62. D. Colton, P. Kress. *Integral equation methods in scattering theory.* John Wiley & Sons, New York, (1983).

63. V.P. Mikhailov. *Partial differential equations.* Nauka, Moscow, (1983) (in Russian).

64. H. Triebel. *Interpolation theory. Function spaces. Differential operators.* Veb Deutshe Verlag, Berlin, (1978).

65. M. Taylor. *Pseudodifferential operators.* Princeton Univ. Press, Princeton, (1981).

66. G.I. Eskin. *Boundary value problems for elliptic pseudodifferential equations.* Nauka, Moscow, (1973) (in Russian).

67. F.G. Ramm. *Scattering by obstacles.* D. Reidel Publ. Comp., Dordrecht, (1986).

68. E. Stephan, W.L. Wendland. An augmented Galerkin procedure for the boundary integral method applied to two-dimensional screen and crack problems. *Applicable Analysis*, **18**, 105-128 (1984).

69. B.-W. Rempel, S. Schulze. *Index theory of elliptic boundary value problems.* Akademie-Verlag, Berlin, (1982).

70. Yu.V. Egorov, M.A. Shubin. Linear partial differential equations. Elements of morden theory. In.: *Modern problems of mathematics. Fundamental trends.* **31**, 5-125, VINITI, Moscow, (1988) (in Russian).

71. S. Prössdorf. *Einige Klassen singulärer Gleichungen*, Akademie-Verlag, Berlin, (1974).

72. E.V. Chernokozhin. *On the spectrum of an operator-valued function describing the normal waves of a microstrip line*, Available from VINITI, no. 4485-84 DEP, Moscow, (1984) (in Russian).

73. A.I. Guseinov, K.S. Mukhtarov. *Introduction to the theory of nonlinear singular integral equations*, Nauka, Moscow, (1980) (in Russian).

74. L.V. Kantorovich, G.P. Akilov. *Functional analysis*, Nauka, Moscow, (1984) (in Russian).

75. A.S. Ilyinsky, E. V. Chernokozhin. On the regularization of an integral operator with logarithmic singularities of the kernel. *Differents. Uravn.*, **24**, 1433–1437 (1988).

76. L. Lewin. *Theory of waveguides*, Newnes-Butterworths, London, (1975).

77. E. Janke, F. Emde, F. Losch. *Tafeln Höherer Funktionen*, Teubner, Stuttgart, (1960).

78. *Handbook of mathematical functions with formulas, graphs, and mathematical tables*, Ed. by M. Abramovitz, I. Stegun, U.S. Govt. Printing Office, Washington, (1964).